JOURNAL OF ICT STANDARDIZATION

Volume 3, No. 1 (July 2015)

Special Issue on
ITU Kaleidoscope 2014: QoS and Network Crawling

Guest Editors:
Malcolm Johnson
Deputy Secretary-General
International Telecommunication Union
Chaesub Lee
Director of the Telecommunication Standardization Bureau (TSB)
International Telecommunication Union

Guest Associate Editor:
Alessia Magliarditi
Programme Coordinator
Policy and Technology Watch Division, TSB

Special Issue on
SDN/NFV Standardization Activities

Guest Editors:
Martin Stiemerling
University of Applied Sciences Darmstadt IETF Transport Area Director
Marcus Schöller
University of Applied Sciences Reutlingen ETSI NFV REL Chairman

JOURNAL OF ICT STANDARDIZATION

Chairperson: Ramjee Prasad, CTIF, Aalborg University, Denmark
Editor-in-Chief: Anand R. Prasad, NEC, Japan
Advisors: Bilel Jamoussi, ITU, Switzerland
Jesper Jerlang, Dansk Standard, Denmark

Editorial Board
Kiritkumar Lathia, Independent ICT Consultant, UK
Hermann Brandt, ETSI, France
Kohei Satoh, ARIB, Japan
Sunghyun Choi, Seoul National University, South Korea
Ashutosh Dutta, AT&T, USA
Alf Zugenmaier, University of Applied Sciences Munich, Germany
Julien Laganier, Luminate Wireless, Inc., USA
John Buford, Avaya, USA
Monique Morrow, Cisco, Switzerland
Vijay K. Gurbani, Alcatel Lucent, USA
Henk J. de Vries, Rotterdam School of Management,
Erasmus University, The Netherlands
Yoichi Maeda, TTC Japan
Debabrata Das, IIIT-Bangalore, India
Signe Annette Bøgh, Dansk Standard, Denmark
Rajarathnam Chandramouli, Stevens Institute of Technology, USA

Objectives

- Bring papers on new developments, innovations and standards to the readers
- Cover pre-development, including technologies with potential of becoming a standard, as well as developed / deployed standards
- Publish on-going work including work with potential of becoming a standard technology
- Publish papers giving explanation of standardization and innovation process and the link between standardization and innovation
- Publish tutorial type papers giving new comers a understanding of standardization and innovation

Aims

- The aim of this journal is to publish standardized as well as related work making "standards" accessible to a wide public – from practitioners to new comers.
- The journal aims at publishing in-depth as well as overview work including papers discussing standardization process and those helping new comers to understand how standards work.

Scope

- Bring up-to-date information regarding standardization in the field of Information and Communication Technology (ICT) covering all protocol layers and technologies in the field.

JOURNAL OF ICT STANDARDIZATION

Volume 3, No. 1 (July 2015)

Published, sold and distributed by:
River Publishers
Niels Jernes Vej 10
9220 Aalborg Ø
Denmark

River Publishers
Lange Geer 44
2611 PW Delft
The Netherlands

www.riverpublishers.com

Journal of ICT Standardization is published three times a year. Publication programme, 2014–2015: Volume 3 (3 issues)

ISSN: 2245-800X (Print Version)
ISSN: 2246-0853 (Online Version)
ISBN: 978-87-93379-07-7

Guest Editorial for Special Issue on ITU Kaleidoscope 2014: QoS and Network Crawling

The Kaleidoscope academic conference is ITU's flagship academic event. Established in 2008, the conference brings the work of ICT researchers to the attention of the standardization community.

The conference has matured into one of the highlights of ITU's calendar of events.

Kaleidoscope sheds light on research at an early stage in the interests of identifying associated standardization needs. Oriented towards the future, the research findings presented to the conference assist ITU in planning the course of its international standardization work.

The sixth edition of Kaleidoscope in 2014 took the theme "Living in a converged world – impossible without standards?" It was tackled from the variety of perspectives that has become essential in the context of technological and industrial convergence.

From a total of ninety-eight submissions from thirty-nine countries, a double-blind, peer-review process selected thirty-four papers for presentation at the conference, all of which were published in the Kaleidoscope Proceedings as well as the IEEE *Xplore* Digital Library. A selection of the best papers were also published in the March 2015 edition of IEEE Communications Magazine as part of the second issue of the Communications Standards Supplement.

This is the third in a series of three special issues to showcase extended versions of selected Kaleidoscope papers. The first addressed the theme "Towards 5G" and was published in November 2014 (Volume 2, No. 2). The second, published in March 2015 (Volume 2, No. 3), covered "Assessments, Models and Evaluation" to monitor the quality of ICT services.

The first Kaleidoscope paper in this third special issue looks at the calculation of Quality of Service (QoS) parameters for two resource-intensive services – video conferencing and video on demand (VoD) – provided over 4G mobile-wireless networks. The second featured paper proposes a new web-crawling technology to identify university courses that offer education about international standardization, in addition exploring the merits of university collaboration in providing such education.

The two papers and their respective authors are as follows:

1. Algorithm for Calculating QoS Parameters of Video Conferencing and Video on Demand Services in Wireless Next Generation Networks.

Vladimir Y. Borodakiy (JSC "Concern Sistemprom", Moscow, Russia), Konstantin E. Samouylov, Irina A. Gudkova, Ekaterina V. Markova (Department of Applied Probability and Informatics, Peoples' Friendship University of Russia, Moscow, Russia).

2. Syllabuses Crawling and Knowledge Extraction of Courses about Global Standardization Education.

Hiroshi Nakanishi (Osaka University, Japan), Tetsuo Oka (Tresbind Corporation, Japan), Yoshiaki Kanaya (Brain Gate Co. LTD, Japan).

We would like to thank the authors for their preparation of extended papers, the papers' reviewers for their generous contribution of time and expertise, and of course the readers of this journal for their interest and feedback on this third Kaleidoscope special issue.

Readers are also encouraged to participate in the 2015 edition of ITU Kaleidoscope, "Trust in the Information Society", to be held at the Universitat Autònoma de Barcelona, Spain, 9–11 December 2015. For more information on Kaleidoscope 2015, please consult the event's webpage: http://itu.int/go/K-2015.

Guest Editors: Malcolm Johnson
Deputy Secretary-General
International Telecommunication Union

Chaesub Lee
Director of the Telecommunication Standardization
Bureau (TSB), International Telecommunication Union

Guest Associate Editor: Alessia Magliarditi
Programme Coordinator, Policy and Technology
Watch Division, TSB

Algorithm for Calculating QoS Parameters of Video Conferencing and Video on Demand Services in Wireless Next Generation Networks*

Vladimir Y. Borodakiy[1], Konstantin E. Samouylov[2], Irina A. Gudkova[2] and Ekaterina V. Markova[2]

[1]*JSC "Concern Sistemprom", Moscow, Russia*
[2]*Department of Applied Probability and Informatics, Peoples' Friendship University of Russia, Moscow, Russia*
E-mail: bvu@systemprom.ru; {ksam, igudkova, emarkova}@sci.pfu.edu.ru

Received November 2014;
Accepted March 2015

Abstract

The standardization of 5G wireless next generation networks is planned to be started by 3GPP consortium in 2020. Nevertheless, for today, there is the lack of QoS related 3GPP and ITU-T recommendations describing various popular services, for example video conferencing. The problem is to find the optimal bit rate values for this service not affecting the background lower priority services. In this paper, we propose a tool to solve this problem. The tool includes, first, a mathematical model of pre-emption based radio admission control scheme for video conferencing and background video on demand, and second, an effective algorithm for calculating QoS parameters for these services. The results of the paper could be used to find the optimal bit rates for video conferencing and further develop a proposition for standardization activities in video conferencing recommended parameters.

*The reported study was partially supported by RFBR, research project No. 13-07-00953.

Journal of ICT, Vol. 3, 3–28.
doi: 10.13052/jicts2245-800X.311

Keywords: 5G, LTE, video conferencing, video on demand, radio admission control, pre-emption, service degradation, service interruption, QoS, blocking probability, pre-emption probability, mean bit rate, teletraffic theory, queuing theory.

1 Introduction

We perceive emerging developments in ICTs in various areas of our dynamic life. The concept of getting services and applications "at any time" and "in any place" imposes the corresponding development in wireless next generation networks. The rapid worldwide deployment of 4G LTE [1] cellular networks as well as the further its development towards 5G and the increasing demand for new services are the key ICT trends that highlight the need for future improvements in technologies underlying LTE. Although 3GPP (3ed Generation Partnership Project) consortium concentrates all standardization activities in LTE (see TS 36.300 for a description of the LTE architecture), ITU-T recommendations reflect the main quality of service (QoS) requirements in ICTs. Nevertheless, QoS related ITU-T recommendations (i.e. E.800–E.899, G.1000–G.1999, Y.1500–Y.1599, Y.2100–Y.2199 series) do not describe all the variety of products and services provided by cellular networks. This fact raises the need for updates in these recommendations for 4G and further 5G recommendations. In 2020, 3GPP plans to release the first set of specifications that will define 5G networks. Until 2020, 3GPP proposes to conduct scientific research for 5G specifications development. Thus, until 2020, there will be no standardized solutions, methods and technologies for 5G networks. Therefore, an important task is to develop offers for improving the technological base of 4G networks towards 5G.

LTE and LTE-Advanced networks deployment is inseparably linked with maintaining the quality of service (QoS) and enhancing customer base. Meeting the corresponding radio resource related requirements is the primarily aim of the radio resource management (RRM). Managing radio resources encompasses the radio admission control (RAC), dynamic resource allocation, inter-cell interference coordination (ICIC), etc. RAC schemes are closely related to the types of services provided to customers. The 3GPP specifications (TS 36.300, TS 23.401, TS 23.203) for LTE and LTE-Advanced networks specify nine service classes (QoS class identifier, QCI) that differ in terms of the bit rate, priority level, and packet error loss.

Each of the nine service classes could be provided on guaranteed bit rate (GBR) or on non-guaranteed bit rate (non-GBR). The first four classes are GBR services, for example, video on demand (VoD), other five classes are non-GBR services, for example, web browsing, email. GBR services could be provided not only on one value of bit rate. Its bit rate can change from a maximum value – the so-called maximum bit rate (MBR) – to a minimum value – GBR – depending on the cell load and RAC scheme. This bit rate changes do not alter the service duration. Whereas, the bit rate changes for non-GBR services result in varying the service duration. According to these principles, in terms of teletraffic and queuing theories, services provided on GBR correspond to streaming traffic, services provided without GBR correspond to elastic traffic. In accordance with two communication technologies point-to-point and point-to-multipoint streaming traffic is divided into two subtypes – unicast and multicast traffic. Unlike unicast traffic, multicast traffic has a network resources saving nature, which is achieved through employing multicast technology. So, overall LTE traffic may be divided into three types: unicast streaming, multicast streaming, and elastic traffics.

Each of service classes is associated with a key attribute called the allocation and retention priority (ARP). The value of ARP is used by the RAC as a flag for admitting or rejecting requests of users for service providing. According to 3GPP TS 23.203, "The range of the ARP priority level is 1 to 9 with 1 as the highest level of priority. The pre-emption capability information defines whether a service data flow can get resources that were already assigned to another service data flow with a lower priority level. The pre-emption vulnerability information defines whether a service data flow can lose the resources assigned to it in order to admit a service data flow with a higher priority level." In accordance with this definition ARP contains three information fields, namely, priority value, pre-emption capability and pre-emption vulnerability. The priority value is used for differentiation purposes and it ensures that the request for service with a higher priority level will be accepted. Note that the highest priority, which is equal to one, has signalling traffic, followed by GBR services – priority values from two to five – and the last non-GBR services have the lowest priority from the sixth to the ninth.

From the above 3GPP definition of "pre-emption", it is evident that during the lack of radio resources, assigning it to lower priority services could be realized, at best, through the service degradation, which is also referred to as bandwidth adaptation or partial pre-emption, or, at worst, through the service interruption, which is also referred to as cut off process or full pre-emption. The 3GPP specifications do not specify RAC schemes, and operators have to develop and select an optimal scheme accounting for the service level agreement. Researchers have proposed various RAC schemes [2–6] with different approaches to pre-empting. Nevertheless, the basic principle of service degradation and interrupting holds.

The admission control is realized on the bit rate basis [2–4] or on the cell load basis [5, 6]. Pre-emption algorithms optimize some objective function, i.e. maximize the number of users in a cell [2] or minimize the number of users perceiving service degradation [3]. In turn, the service degradation could be specific to a particular service class [4] or egalitarian to them [5]. Service interrupting generally goes with service degradation and represents the so-called second phase of pre-empting [4, 5]. The performance analysis of RAC schemes requires mathematical methods, primarily, mathematical teletraffic and queuing theories [7]. These methods are widely applied for modelling and analysing not only the last mentioned RAC problem [8, 9], but also dynamic resource allocation [10] and ICIC [11] problems.

The remainder of this paper is organized as follows. In Section 2, we propose a model of RAC scheme for two GBR services: video conferencing (QCI = 2, multicast multi-rate, higher priority) and video on demand (QCI = 4, unicast, lower priority). In Section 3, we derive a recursive algorithm for calculating model QoS parameters. In Section 4, we conduct an analysis of model QoS parameters. Finally, we conclude the paper in Section 5.

2 Markov Model of Radio Admission Control Scheme

2.1 Assumptions and Parameters

We consider a single cell with a total capacity of C bandwidth units (b.u.) supporting two GBR services: multicast video conferencing (VC) service and unicast video on demand (VoD) service. All necessary notations are given in Table 1. The VoD service is provided on single GBR d b.u. Without loss of generality, we assume $d = 1$ b.u. The VC service is a multi-rate service, i.e. its bit rate can be adaptively changed from a maximum value of b_1 b.u. to a minimum value of b_K b.u. according to a given set of values $b_1 > \ldots > b_k > \ldots > b_K$ that depends on the cell load expressed in the number of users.

Let arrival rates λ [1/time unit] (VC) and v [1/time unit] (VoD) be Poisson distributed and let the service time be exponentially distributed with means μ^{-1} [1/time unit] (VC) and κ^{-1} [1/time unit] (VoD). Then we denote the corresponding offered loads as $\rho = \lambda/\mu$ and $a = v/\kappa$.

Table 1   Parameters

Notation	Parameter
C	Downlink peak bit rate, bps
	VC service
$b_1 > \ldots > b_k > \ldots > b_K$	Bit rates for VC service, bps
λ	Arrival rate of requests for VC service, 1/s
μ^{-1}	VC service time, s
$\rho = \lambda/\mu$	VC offered load
$m_k = 1$	Multicast session is active – VC service is provided at least to one user on bit rate b_k
$m_k = 0$	Multicast session is not active – VC service is not provided to users on bit rate b_k
$\mathbf{m} = (m_1, \ldots, m_k, \ldots, m_K)$	State of a multicast VC session
$b(\mathbf{m})$	Bit rate for VC, when the state of a multicast session is $\mathbf{m}$
	VoD service
$d = 1$	Bit rate for VoD service, bps
v	Arrival rate of requests for VoD service, 1/s
κ^{-1}	VoD service time, s
$a = v/\kappa$	VC offered load
n	Number of VoD users
$(\mathbf{m}, n)$	State of the system
$c(\mathbf{m}, n)$	Capacity occupied, when the system is in state $(\mathbf{m}, n)$

Let us introduce the following notations:

- $n \in \{0, 1, \ldots, \lfloor C/d \rfloor\} = \{0, 1, \ldots, C\}$ – number of VoD users ($d = 1$ b.u.);
- $\mathbf{m} = (m_1, \ldots, m_K)$ – state of a multicast session, where m_k can be equal to 1 if session is active on bit rate b_k, i.e. multicast VC service is provided at least to one user on bit rate b_k, or m_k can be equal to 0 if the session is not active on bit rate b_k, i.e. multicast VC service is not provided to users on bit rate b_k, $k = 1, \ldots, K$;
- $(\mathbf{m}, n)$ – state of the system;
- $b(\mathbf{m})$ – bit rate for VC service, when the state of a multicast session is $\mathbf{m}$,

$$b(\mathbf{m}) = \begin{cases} 0, & \text{if } \mathbf{m} = \mathbf{0}, \\ b_k, & \text{if } \mathbf{m} = \mathbf{e}_k, \quad k = 1, \ldots, K; \end{cases}$$

- $c(\mathbf{m}, n) = b(\mathbf{m}) + n$ – capacity occupied, when the system is in state $(\mathbf{m}, n)$.

2.2 RAC Scheme

The VC priority level is higher than the VoD one. First, this fact is realized by the adaptive change of VC bit rate. Second, the RAC is achieved in the way that a new VC request is accepted by the so-called pre-emption owing to the lack of free radio resources. Pre-empting refers to the release of radio resources occupied by VoD service (Table 2, Figure 1). VoD users to be interrupted are selected randomly.

Given the above considerations, when a new VC request arrives, two scenarios are possible.

- The VC request will be accepted on bit rate b_k and the number of VoD users will not be changed, which is possible if the request finds the cell having greater than or equal to b_k b.u. free, $k = 1, \ldots, K$.

Table 2   Fields of ARP for VC and VoD services

	Pre-emption capable	Pre-emption vulnerable
VC (multicast)	Yes (interrupt VoD)	Yes (degraded by VoD)
VoD (unicast)	Yes (degrade VC)	Yes (interrupted by VC)

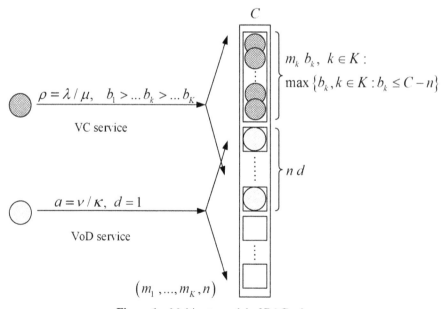

Figure 1   Multi-rate model of RAC scheme.

- The VC request will be accepted on bit rate b_K and $b_K - (C - n)$ VoD users will be pre-empted, which is possible if the request finds the cell having less than b_K b.u. free and n VoD users.

Similarly, when a new VoD request arrives, three scenarios are possible.

- The VoD request will be accepted on bit rate $d = 1$ b.u. without any effect on VC users, which is possible if the request finds the cell having greater than or equal to $d = 1$ b.u. free.
- The VoD request will be accepted on bit rate $d = 1$ b.u. with degrading VC service, which is possible if the request finds the cell having less than $d = 1$ b.u. free and VC service is provided at least to one user on bit rate b_k b.u., $k = 1, \ldots, K - 1$.
- Otherwise, the VoD request will be blocked without any after-effect on the corresponding Poisson process arrival rate.

2.3 Example of Pre-emption Mechanism

We comment the main principles of model functioning – service interruption – with the aid of the following example (Figure 2). Consider a single cell with a total capacity of $C = 5$ b.u. supporting VC and VoD services. The bit rates are $b_1 = 3$ b.u., $K = 1$ and $d = 1$ b.u.

At time t_0 VC service is provided at least to one user, VoD service is provided to two users and the cell has no free resources. At time t_1 the

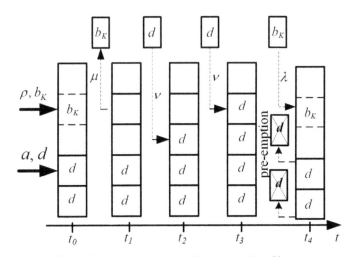

Figure 2 Pre-emption – service interruption $K = 1$.

providing of VC service is finished and the cell has three b.u. free. At time t_2 and t_3 arrived VoD requests are accepted for service and the cell has one b.u. free. Finally, at time t_4 an arrived VC request is accepted for service due to pre-empting of two VoD users.

We do not give an example for the service degradation mechanism due to its simplicity.

2.4 QoS Parameters

According to the above considerations, we denote the system state space as

$$\mathbf{X} = \{(\mathbf{m}, n) : \mathbf{m} = 0, \ n = 0, \ldots, C, \mathbf{m} = \mathbf{e}_1, n = 0, \ldots, C - b_1,$$
$$\mathbf{m} = \mathbf{e}_k, n = C - b_{k-1} + 1, \ldots, C - b_k, k = 2, \ldots, K\},$$

where $\mathbf{e}_k = \left(0, \ldots, 0, \overset{k}{1}, 0, \ldots, 0\right)$.

The process representing the system states is described by the state transition diagram and systems of equilibrium equations, which one can find in the Appendix, A.1. Having found probability distribution $p(\mathbf{m}, n)$, $(\mathbf{m}, n) \in \mathbf{X}$ of the multi-rate model of the RAC scheme for VC and VoD services, one may compute its QoS parameters, notable:

- blocking probability for VoD service

$$B = \sum_{(\mathbf{m},n) \,\in\, \mathbf{X}:\ c(\mathbf{m},n)=C} p(\mathbf{m}, n); \qquad (1)$$

- pre-emption probability for VoD service

$$\Pi = \sum_{(\mathbf{0},n) \,\in\, \mathbf{X}:\ n > C - b_K} \Pi_n \cdot p(\mathbf{0}, n), \qquad (2)$$

$$\Pi_n = \begin{cases} \dfrac{\lambda}{\lambda + \nu + n\kappa} \cdot \dfrac{\dbinom{n-1}{b_K - (C - n) - 1}}{\dbinom{n}{b_K - (C - n)}}, & n = C - b_K + 1, \ldots, C - 1, \\[2em] \dfrac{\lambda}{\lambda + C\kappa} \cdot \dfrac{\dbinom{C-1}{b_K - 1}}{\dbinom{C}{b_K}}, & n = C; \end{cases}$$

- mean bit rate for VC service

$$\overline{b} = \sum_{(\mathbf{m},n) \in \mathbf{X} \,:\, \mathbf{m} \neq \mathbf{0}} b(\mathbf{m}) \cdot \frac{p(\mathbf{m}, n)}{\sum_{(\tilde{\mathbf{m}},\tilde{n}) \in \mathbf{X} \,:\, \tilde{\mathbf{m}} \neq \mathbf{0}} p(\tilde{\mathbf{m}}, \tilde{n})}; \qquad (3)$$

- utilization factor of the cell, i.e. mean load per bandwidth unit

$$\text{UTIL} \cdot C = \sum_{(\mathbf{m}, n) \in \mathbf{X}} c(\mathbf{m}, n) \cdot p(\mathbf{m}, n). \qquad (4)$$

More detailed formulas for calculating model QoS parameters are presented in Appendix, A.2. It seems only possible to determine the system probability distribution by means of numerical methods for solving systems of equilibrium equations.

3 Algorithm for Calculating QoS Parameters

Because the solution of the system of equilibrium equations for the model described above is time-consuming, we consider a simplified model – stochastic equivalent model, the transition to which is described in Appendix, A.3. We note only that in a simplified model VC bit rates are not considered to form the system state. The state of multicast session m is only important, which can be equal to 1 if session is active, or can be equal to 0 if the session is not active.

It could be proved that the process representing the system states is not a reversible Markov process, and to determine the system probability distribution $P(m, n)$ we need to get a recursive algorithm.

Algorithm. Step 1.1. Determine the coefficients $\alpha_{00}, \beta_{00}, \alpha_{10}, \beta_{10}$ from the relations

$$\alpha_{00} = 1, \quad \beta_{00} = 0, \quad \alpha_{10} = 0, \quad \beta_{10} = 1.$$

Step 1.2. Calculate the coefficients $\alpha_{01}, \beta_{01}, \alpha_{11}, \beta_{11}$ by formulas

$$\alpha_{01} = \frac{v + \lambda}{\kappa}, \quad \beta_{01} = -\frac{\mu}{\kappa}, \quad \alpha_{11} = -\frac{\lambda}{\kappa}, \quad \beta_{11} = \frac{v + \mu}{\kappa}.$$

Step 1.3. Calculate the coefficients $\alpha_{0n}, \beta_{0n}, \alpha_{1n}, \beta_{1n}, n = 2, \ldots, C - b_K$ by formulas

$$\alpha_{0n} = \frac{1}{n} \left((\alpha_{01} + (n - 1)) \alpha_{0,n-1} + \beta_{01} \alpha_{1,n-1} - a \alpha_{0,n-2} \right),$$

$$\beta_{0n} = \frac{1}{n} \left((\alpha_{01} + (n-1)) \beta_{0,n-1} + \beta_{01} \beta_{1,n-1} - a\beta_{0,n-2} \right),$$

$$\alpha_{1n} = \frac{1}{n} \left((\beta_{11} + (n-1)) \alpha_{1,n-1} + \alpha_{11} \alpha_{0,n-1} - a\alpha_{1,n-2} \right),$$

$$\beta_{1n} = \frac{1}{n} \left((\beta_{11} + (n-1)) \beta_{1,n-1} + \alpha_{11} \beta_{0,n-1} - a\beta_{1,n-2} \right).$$

Step 1.4. Calculate the coefficients $\alpha_{0,C-b_K+1}$, $\beta_{0,C-b_K+1}$ by formulas

$$\alpha_{0,C-b_K+1} = \frac{1}{C-b_K+1} \left((\alpha_{01} + (C-b_K))\alpha_{0,C-b_K} \right.$$
$$+ \left. \beta_{01}\alpha_{1,C-b_K} - a\alpha_{0,C-b_K-1} \right),$$

$$\beta_{0,C-b_K-1} = \frac{1}{C-b_K+1} \left((\alpha_{01} + (C-b_K))\beta_{0,C-b_K} \right.$$
$$+ \left. \beta_{01}\beta_{1,C-b_K} - a\beta_{0,C-b_K-1} \right).$$

Step 1.5. Calculate the coefficients α_{0n}, β_{0n}, $n = C - b_K + 2, \ldots, C$ by formulas

$$\alpha_{0n} = \frac{1}{n} \left((\alpha_{01} + (n-1)) \alpha_{0,n-1} - a\alpha_{0,n-2} \right),$$

$$\beta_{0n} = \frac{1}{n} \left((\alpha_{01} + (n-1)) \beta_{0,n-1} - a\beta_{0,n-2} \right).$$

Step 2.1. Calculate the value of variable x by solving the global balance equations for the boundary state $(0, C)$ (Appendix, A.3) by the formula

$$x = \frac{\frac{\nu}{\lambda + C\kappa}\alpha_{0,C-1} - \alpha_{0,C}}{\beta_{0,C} - \frac{\nu}{\lambda + C\kappa}\beta_{0,C-1}}.$$

Step 2.2. Calculate the value of unnormalized probabilities

$$q(m,n) = \alpha_{mn} + \beta_{mn} \cdot x, \quad m = 0, \ n = 0,.., C \ \vee \ m = 1,$$
$$n = 0,.., C - b_K.$$

Step 3.1. Calculate the normalizing factor

$$G = \sum_{\substack{(i,j)\,:\,i=0,\ j=0,..,C\ \vee \\ i=1,\ j=0,..,C-b_K}} q(i,j).$$

Step 3.2. Calculate the probability distribution

$$P(m,n) = \frac{q(m,n)}{G}, \ m = 0, \ n = 0,.., C \ \lor \ m = 1, n = 0,.., C - b_K.$$

Step 4.1. Calculate the blocking probability

$$B = P(0,C) + P(1, C - b_K). \tag{5}$$

Step 4.2. Calculate the pre-emption probability

$$\Pi = \sum_{n=C-b_K+1}^{C-1} \frac{\lambda}{\lambda + v + n\kappa} \frac{b_K - C + n}{n} P(0,n)$$

$$+ \frac{\lambda}{\lambda + C\kappa} \frac{b_K}{C} P(0,C). \tag{6}$$

Step 4.3. Calculate the mean bit rate

$$\bar{b} = \frac{b_1 \sum_{n=0}^{C-b_1} P(1,n) + \sum_{k=2}^{K} b_k \sum_{n=C-b_{k-1}+1}^{C-b_k} P(1,n)}{\sum_{n=0}^{C-b_K} P(1,n)}. \tag{7}$$

Step 4.5. Calculate the utilization factor

$$\text{UTIL} \cdot C = \sum_{n=1}^{C} nP(0,n) + \sum_{n=0}^{C-b_1} (b_1 + n) P(1,n)$$

$$+ \sum_{k=2}^{K} \sum_{n=C-b_{k-1}+1}^{C-b_k} (b_k + n) P(1,n). \tag{8}$$

4 Analysing QoS Parameters of Video Conferencing and Video on Demand

4.1 Numerical Example

According to forecasts by Cisco Systems [12], mobile applications providing video services will generate most of global mobile data traffic by 2018, namely about 69 percent. However, not only in the future, but also in the beginning of 2012, mobile video represents more than half of global mobile data traffic.

In this regard, let us consider an example of a single cell supporting VC and VoD services to illustrate the performance measures defined above. One of the main trends of 5G networks is the need to save resources to allocate for high quality video services. This is possible through the implementation of an individual functional element of the network that operates on the basis of the multicast technology. Note that the users of LTE networks got the possibility to consume multicast services with the inclusion of MBMS (Multimedia Broadcast and Multicast Service) subsystem in SAE (System Architecture Evolution).

We numerically analyse the effect of increasing the popularity of multicast service (i.e. video conferencing) and corresponding changes in the QoS parameters of unicast service (i.e. video on demand). Thus, the performance analysis of the pre-emption based RAC scheme for VC and VoD services is performed under the conditions of different VC popularity expressed in the ratio α of VC arrival rate to the total arrival rate. To give you the numerical example, we will consider the requirements of Skype, which are recommended for providing VC services to a group from seven and more people [13].

Let us write out the input data of the numerical example in the Table 3. We compute blocking probability B (2.2, 3.2, Figure 3), pre-emption probability Π (2.3, 3.3, Figure 4), mean bit rate $\overline{b}$ (2.4, 3.4, Figure 5), and utilization factor UTIL (2.5, 3.5, Figure 6), by changing the offered load $\rho + a$.

The growth of the cell load results in increasing blocking probability, pre-emption probability and utilization factor (Figure 3, 5, 6), but in decreasing mean bit rate (Figure 4). The numerical example (Figure 4) shows that the greater multicast service popularity, the greater is its mean bit rate. This is due to the fact that the increasing popularity of multicast services is expressed in the fact that the multicast session is active most of time, and interruptions of unicast service users are less frequent – which is confirmed by Figure 3. Therefore, the increasing popularity of multicast services does not result in significant deterioration of the QoS parameters of unicast service.

Table 3   Numerical data

VC (multicast traffic)		VoD (unicast traffic)	
$C = 50$ Mbps, $\alpha = 0.3,\ \ 0.5,\ \ 0.7,$		$\rho + a = 0 \div 10$	
$\rho = \alpha \cdot (\rho + a),$		$a = (1 - \alpha) \cdot (\rho + a),$	
$\mu^{-1} = 1$ hour, $\lambda = \rho\mu,$		$\kappa^{-1} = 2$ hours,	
$b_1 = 8$ Mbps, $b_2 = 6$ Mbps, $b_3 = 4$ Mbps		$v = a\kappa,\ \ d = 2$ Mbps	

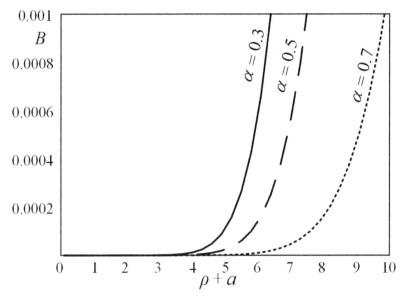

Figure 3 Blocking probability for VoD service ($C = 50$ Mbps, $b_1 = 8$ Mbps, $b_2 = 6$ Mbps, $b_3 = 4$ Mbps, $d = 2$ Mbps, $\mu^{-1} = 1$ hour, $\kappa^{-1} = 2$ hours).

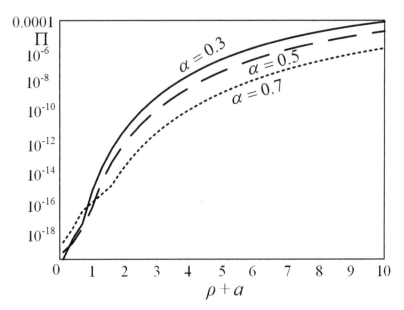

Figure 4 Pre-emption probability for VoD service ($C = 50$ Mbps, $b_1 = 8$ Mbps, $b_2 = 6$ Mbps, $b_3 = 4$ Mbps, $d = 2$ Mbps, $\mu^{-1} = 1$ hour, $\kappa^{-1} = 2$ hours).

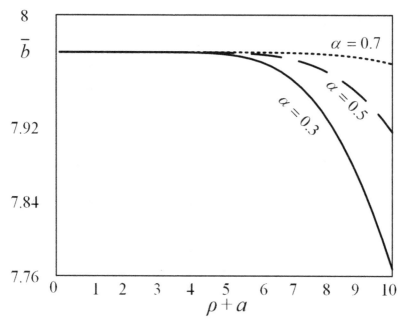

Figure 5   Mean bit rate for VC service ($C = 50$ Mbps, $b_1 = 8$ Mbps, $b_2 = 6$ Mbps, $b_3 = 4$ Mbps, $d = 2$ Mbps, $\mu^{-1} = 1$ hour, $\kappa^{-1} = 2$ hours).

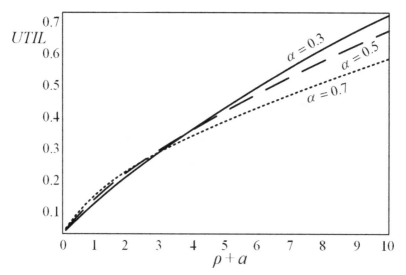

Figure 6   Utilization factor ($C = 50$ Mbps, $b_1 = 8$ Mbps, $b_2 = 6$ Mbps, $b_3 = 4$ Mbps, $d = 2$ Mbps, $\mu^{-1} = 1$ hour, $\kappa^{-1} = 2$ hours).

4.2 Optimization Problem for Video Conferencing Bit Rates

In this paper, we propose a model of the RAC scheme for VC and VoD services, allowing to estimate the following QoS parameters: blocking and pre-emption probabilities for VoD service, mean bit rate for VC service and utilization factor of the cell. To calculate these characteristics a recursive algorithm was proposed. The obtained results allow us to perform numerical experiments to further development of recommendations for 5G networks. These recommendations will provide the opportunity to make a choice of services parameter values depending on the user QoS requirements. To achieve this goal it is necessary to solve the optimization problem. The objective functions of this problem are all listed above QoS parameters. As a solution to this multi-objective problem is time-consuming, its solution is often limited to the optimization of one of the QoS parameters in the constraints on other parameters.

Consider an example of such optimization problem. The task is to maximize the average bit rate $\bar{b}$ of the VC service with the blocking probability and pre-emption probabilities for VoD service not exceeding the values B^* and Π^* respectively, and the bit rate for VC service being not smaller than b^*. The average VC bit rate depends on the initial set $\mathbf{D} = \{d_1, d_2, \ldots, d_M\}$ of values and their number K. Then the problem of VC bit rate optimization can be formulated as follows.

$$\bar{b}(K, b_1, \ldots, b_K) \to \max,$$

$$\text{VC}: \begin{cases} b_k \in \mathbf{D} = \{d_1, d_2, \ldots, d_M\}, \ k = 1, \ldots, K, \\ b_1 > b_2 > \ldots > b_K, \\ b_K \geq b^*, \end{cases}$$

$$\text{VoD}: \begin{cases} B(K, b_1, \ldots, b_K) \leq B^*, \\ \Pi(K, b_1, \ldots, b_K) \leq \Pi^*. \end{cases}$$

Solving the numerical optimization problem for VC and VoD services, we can find the recommended values of bit rates for VC service depending on different values of parameters b^*, B^* and Π^*.

5 Conclusion

In this paper, we addressed an admission control problem for a multi-service LTE radio network, and presented a new multi-rate model for two guaranteed bit rate services: video conferencing and video on demand. The RAC scheme is based on the VC quality degradation from a high to standard definition. The scheme assumes that a VC user can pre-empt high definition VoD users. Considering that the process representing the system states is not a reversible Markov process, we propose the recursive algorithm to calculate the system probability distribution, which is used to analyze the main model QoS parameters.

The results of numerical analysis can be used to plan the standards for 5G wireless cellular networks. They can be used to develop the software that is responsible for radio resources management, namely to develop RAC schemes. An interestigtask for future studies is the development of a simulation model for verification the accuracy of the results.

Appendix

A.1. State Transition Diagram and System of Equilibrium Equations

$$
\left(\lambda \cdot 1\left\{ \mathbf{m} = \mathbf{0}, n \leq C - b_1 \right\} + \lambda \cdot \sum_{k=2}^{K} 1\left\{ \mathbf{m} = \mathbf{0}, C - b_{k-1} < n \leq C - b_k \right\} \right.
$$

$$
+ \lambda \cdot 1\left\{ \mathbf{m} = \mathbf{0}, n > C - b_K \right\} + \nu \cdot 1\left\{ \mathbf{m} = \mathbf{0}, n < C \right\}
$$

$$
+ \nu \cdot \sum_{k=1}^{K} 1\left\{ \mathbf{m} = \mathbf{e}_k, n < C - b_k \right\} + \nu \cdot \sum_{k=1}^{K-1} 1\left\{ \mathbf{m} = \mathbf{e}_k, n = C - b_k \right\}
$$

$$
+ \mu \cdot \sum_{k=1}^{K} 1\left\{ \mathbf{m} = \mathbf{e}_k \right\} + n\kappa \cdot 1\left\{ \mathbf{m} = \mathbf{0}, n > 0 \right\}
$$

$$
\left. + n\kappa \cdot \sum_{k=1}^{K} 1\left\{ \mathbf{m} = \mathbf{e}_k, n > 0 \right\} \right) p\left(\mathbf{m}, n \right)
$$

$$
= \mu p\left(\mathbf{e}_1, n \right) \cdot 1\left\{ \mathbf{m} = \mathbf{0}, n \leq C - b_1 \right\}
$$

$$
+ \mu \sum_{k=2}^{K} p\left(\mathbf{e}_k, n \right) \cdot 1\left\{ \mathbf{m} = \mathbf{0}, C - b_{k-1} < n \leq C - b_k \right\} +
$$

$$
+ \left(n + 1 \right) \kappa p\left(\mathbf{0}, n + 1 \right) \cdot 1\left\{ \mathbf{m} = \mathbf{0}, n < C \right\}
$$

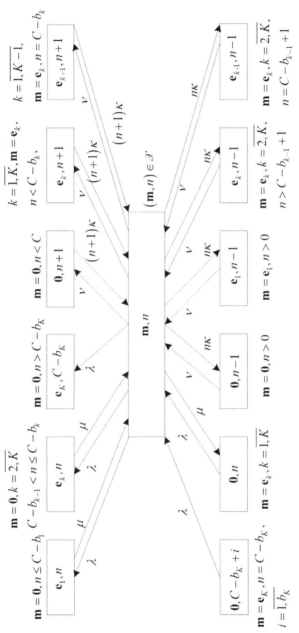

Figure A.1 State transition diagram for multi-rate model.

$$+ (n+1)\,\kappa \sum_{k=1}^{k} p\left(\mathbf{e}_k, n+1\right) \cdot 1\left\{\mathbf{m} = \mathbf{e}_k, n < C - b_k\right\} +$$

$$+ (n+1)\kappa \sum_{k=1}^{k-1} p(\mathbf{e}_{k+1}, n+1) \cdot \left\{\mathbf{m} = \mathbf{e}_k, n = C - b_k\right\}$$

$$+ \lambda \sum_{k=1}^{K} p\left(\mathbf{0}, n\right) \cdot 1\left\{\mathbf{m} = \mathbf{e}_k\right\} +$$

$$+ \lambda \cdot 1\left\{\mathbf{m} = \mathbf{e}_K, n = C - b_K\right\} \sum_{i=1}^{b_K} p\left(\mathbf{0}, C - b_K + i\right)$$

$$+ \nu p\left(\mathbf{0}, n-1\right) \cdot 1\left\{\mathbf{m} = \mathbf{0}, n > 0\right\} + +\nu p\left(\mathbf{e}_1, n-1\right) \cdot 1\left\{\mathbf{m} = \mathbf{e_1}, n > 0\right\}$$

$$+ \nu \sum_{k=2}^{K} p\left(\mathbf{e}_k, n-1\right) \cdot 1\left\{\mathbf{m} = \mathbf{e}_k, n > C - b_{k-1} + 1\right\} +$$

$$+ \nu \sum_{k=2}^{K} p\left(\mathbf{e}_{k-1}, n-1\right) \cdot 1\left\{\mathbf{m} = \mathbf{e}_k, n = C - b_{k-1} + 1\right\}, (\mathbf{m}, n) \in \mathbf{X}.$$

A.2. Detailed Formulas for Calculation of Model QoS Parameters

- Formula for calculation of blocking probability (2.2)

$$B = p\left(\mathbf{0}, C\right) + p\left(\mathbf{e}_K, C - b_K\right).$$

- Formula for calculation of pre-emption probability (2.3) for VoD service

$$\Pi = \sum_{n=C-b_K+1}^{C-1} \left[\left(\begin{array}{c} n-1 \\ b_K - (C-n) - 1 \end{array} \right) \middle/ \left(\begin{array}{c} n \\ b_K - (C-n) \end{array} \right) \right]$$

$$\frac{\lambda}{\lambda + \nu + n\kappa} p\left(\mathbf{0}, n\right) + \left[\left(\begin{array}{c} C-1 \\ b_K - 1 \end{array} \right) \middle/ \left(\begin{array}{c} C \\ b_K \end{array} \right) \right] \frac{\lambda}{\lambda + C\kappa} p\left(\mathbf{0}, C\right).$$

- Formula for calculation of mean bit rate (2.4) for VC service

$$\overline{b} = \frac{b_1 \sum\limits_{n=0}^{C-b_1} p\left(\mathbf{e}_1, n\right) + \sum\limits_{k=2}^{K} b_k \sum\limits_{n=C-b_{k-1}+1}^{C-b_k} p\left(\mathbf{e}_k, n\right)}{\sum\limits_{n=0}^{C-b_1} p\left(\mathbf{e}_1, n\right) + \sum\limits_{k=2}^{K} \sum\limits_{n=C-b_{k-1}+1}^{C-b_k} p\left(\mathbf{e}_k, n\right)}.$$

- Formula for calculation of utilization factor UTIL (2.5) of the cell

$$\text{UTIL} \cdot C = \sum_{n=1}^{C} n \cdot p\left(\mathbf{0}, n\right) + \sum_{n=0}^{C-b_1} (b_1 + n) \cdot p\left(\mathbf{e}_1, n\right)$$

$$+ \sum_{k=2}^{K} \sum_{n=C-b_{k-1}+1}^{C-b_k} (b_k + n) \cdot p\left(\mathbf{e}_k, n\right).$$

A.3. Stochastic Equivalent Model

Let us divide all the states of the system in two groups. The first group includes all states in which a multicast session is "on", the second group includes all states in which a multicast session is "off". Since the state of a multicast session can be equal to 0 or 1, the number of VoD users allows to uniquely determine a bit rate b_k, $k \in K$: $\max\{b_k, k \in K : b_k \leq C - n\}$, and the following conformities could be established

$$m\left(\mathbf{m}\right) = \begin{cases} 0, & \text{if } \mathbf{m} = \mathbf{0}, \\ 1, & \text{if } \mathbf{m} = \mathbf{e}_k, \quad k = 1, \ldots, K; \end{cases}$$

or

$$m\left(\mathbf{m}\right) = \begin{cases} \mathbf{0}, & \text{if } m = 0, \ 0 \leq n \leq C; \\ \mathbf{e}_1, & \text{if } m = 1, \ 0 \leq n \leq C - b_1; \\ \mathbf{e}_k, & \text{if } m = 1, \ C - b_{k-1} < n \leq C - b_k, \quad k = 2, \ldots, K. \end{cases}$$

We simplify the model transforming it to a stochastic equivalent model (Figure 1) with the state of the system (m, n). The system state space (2.1) transforms to

$$\mathbf{Y} = \{(0, n), \quad n : (\mathbf{0}, n) \in \mathbf{X}, (1, n),$$
$$n : (\mathbf{e}_k, n) \in \mathbf{X}, \quad k = 1, \ldots, K\}.$$

The system of equilibrium equations is as follows

$$P(0,0)\left(\lambda + v\right) = P(0,1)\kappa + P(1,0)\mu;$$

$$P(1,0)\left(\mu + v\right) = P(0,0)\lambda + P(1,1)\kappa;$$

$$P(0,n)\left(v + \lambda + n\kappa\right) = P(0, n+1)(n+1)\kappa + P(0, n-1)v + P(1,n)\mu,$$
$$n = 1, \ldots, C - b_K;$$

$$P(1,n)\,(v+\mu+n\kappa) = P(1,n+1)(n+1)\kappa + P(0,n)\lambda + P(1,n-1)v,$$
$$n = 1,\ldots,C-b_K-1;$$

$$P(1,C-b_K)\,(\mu+(C-b_K)\kappa) = P(1,C-b_K-1)v + \lambda \sum_{n=C-b_K}^{C} P(0,n);$$

$$P(0,n)\,(v+\lambda+n\kappa) = P(0,n+1)(n+1)\kappa + P(0,n-1)v,$$
$$n = C-b_K+1,\ldots,C-1;$$

$$P(0,C)\,(\lambda+C\kappa) = P(0,C-1)v.$$

To derive the recursive algorithm we use the approach based on partitioning of the system state space. The state space **Y** is partitioned for number n of VoD users

$$\mathbf{Y} = \{(m,n) \in \mathbf{Y} : n = 0\} \cup \{(m,n) \in \mathbf{Y} : n > 0\}.$$

The first space includes the states through which we express the probabilities of states of the second space.

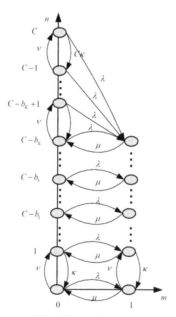

Figure A.3   State transition diagram for stochastic equivalent model.

List of Abbreviations

ARP	Allocation and Retention Priority
b.u.	bandwidth units
GBR	Guaranteed Bit Rate
ICIC	Inter-Cell Interference Coordination
ICT	Information and Communication Technology
MBMS	Multimedia Multicast Broadcast Service
MBR	Maximum Bit Rate
non-GBR	non-Guaranteed Bit Rate
QCI	QoS Class Identifier
QoS	Quality of Service
RAC	Radio Admission Control
RRM	Radio Resource Management
SAE	System Architecture Evolution
SLA	Service Level Agreement
VC	Video Conferencing
VoD	Video on Demand

References

[1] M. Stasiak, M. Glabowski, A. Wisniewski, and P. Zwierzykowski, 'Modelling and dimensioning of mobile wireless networks: from GSM to LTE', Willey, pp. 1–340, 2010.

[2] M. Qian, Y. Huang, J. Shi, Y. Yuan, L. Tian, and E. Dutkiewicz, 'A novel radio admission control scheme for multiclass services in LTE systems', Proc. of the 7th IEEE Global Telecommunications Conference GLOBECOM-2009 (November 30 – December 4, 2009, Honolulu, Hawaii, USA), IEEE, pp. 1–6, 2009.

[3] N. Nasser and H. Hassanein, 'Combined admission control algorithm and bandwidth adaptation algorithm in multimedia cellular networks for QoS provisioning', Proc. of the 17th Canadian Conference on Electrical and Computer Engineering CCECE-2004 (May 2–5, 2004, Niagara Falls, Ontario, Canada), IEEE, vol.2, pp. 1183–1186, 2004.

[4] M. Khabazian, O. Kubbar, and H. Hassanein, 'A fairness-based pre-emption algorithm for LTE-Advanced', Proc. of the 10th IEEE Global Telecommunications Conference GLOBECOM-2012 (December 3–7, 2012, Anaheim, California, USA), IEEE, pp. 5320–5325, 2012.

[5] R. Kwan, R. Arnott, R. Trivisonno, and M. Kubota, 'On pre-emption and congestion control for LTE systems', Proc. of the 72nd Vehicular Technology Conference VTC2010-Fall (September 6–9, 2010, Ottawa, Canada), IEEE, pp. 1–5, 2010.

[6] A.M. Rashwan, A.-E.M. Taha, and H.S. Hassanein, 'Considerations for bandwidth adaptation mechanisms in wireless networks', Proc. of the 24th Biennial Symposium on Communications QBSC-2008 (June 24–26, 2008, Kingston, Ontario, Canada), IEEE, pp. 43–47, 2008.

[7] G.P. Basharin, Y.V. Gaidamaka, and K.E. Samouylov, 'Mathematical theory of teletraffic and its application to the analysis of multiservice communication of next generation networks', Automatic Control and Computer Sciences, vol. 47 no.2, pp. 62–69, 2013.

[8] K.E. Samouylov and I.A. Gudkova, 'Analysis of an admission model in a fourth generation mobile network with triple play traffic', Automatic Control and Computer Sciences, vol. 47 no. 4, pp. 202–210, 2013.

[9] I.A. Gudkova and K.E. Samouylov, 'Modelling a radio admission control scheme for video telephony service in wireless networks', Lecture Notes in Computer Science, vol. 7469, pp. 208–215, 2012.

[10] V.Y. Borodakiy, I.A. Buturlin, I.A. Gudkova, and K.E. Samouylov, 'Modelling and analysing a dynamic resource allocation scheme for M2M traffic in LTE networks', Lecture Notes in Computer Science, vol. 8121, pp. 420–426, 2013.

[11] K.E. Samouylov, I.A. Gudkova, and N.D. Maslovskaya, 'A model for analysing impact of frequency reuse on inter-cell interference in LTE network', Proc. of the 4th International Congress on Ultra Modern Telecommunications and Control Systems ICUMT-2012 (October 3–5, metricconverterProductID2012, St2012, St. Petersburg, Russia), IEEE, pp. 298–301, 2012.

[12] 'Cisco visual networking index: Global Mobile Data Traffic Forecast Update, 2013–2018: usage: White paper', Cisco Systems, 40 p., 2014.

[13] 'How much bandwidth does Skype need?', Skype, 2014. (https://support.skype.com/en/faq/FA1417/how-much-bandwidth-does-skype-need)

Biographies

V. Y. Borodakiy graduated from the National Research Nuclear University "Moscow Engineering Physics Institute" in 2006. In 2009, he received his Ph.D. degree in telecommunication systems and computer networks from the NRNU MEPhI. Vladimir Borodakiy works at JSC "Concern Sistemprom" involved in developing complex distributed control systems and special-purpose communications with the required level of security. His current research interests lie in the area of performance analysis of 4G and 5G networks and industrial control systems. In these areas, he has published several papers in refereed journals and conference proceedings.

K. E. Samouylov received his Ph.D. degree from the Moscow State University and Doctor of Sciences degree from the Moscow Technical University of Communications and Informatics. During 1985–1996, he held several positions at the Faculty of Sciences of the Peoples' Friendship University of Russia where he became a head of the Telecommunication Systems Department in 1996. From 2014, he became a head of the Department of Applied Probability and Informatics of PFUR. During last two decades, Konstantin Samouylov has been conducting research projects for the Helsinki and Lappeenranta Universities of Technology, Moscow Central Science Research Telecommunication Institute, several Institutes of Russian Academy of Sciences and a number of Russian network operators. His current research interests are performance analysis of 4G networks (LTE, WiMAX), teletraffic of triple play networks, signalling network (SIP) planning, and cloud computing. He has written more than 150 scientific and technical papers and three books.

I. A. Gudkova received her B.Sc. and M.Sc. degrees in applied mathematics from the Peoples' Friendship University of Russia in 2007 and 2009, respectively. In 2007, she was awarded a scholarship of the Government of the Russian Federation. In 2011, she received her Ph.D. degree in applied mathematics and computer sciences from the PFUR. Since 2008, Irina Gudkova works at the Telecommunication Systems Department of PFUR, now she is an Associate Professor at the Department of Applied Probability and Informatics of PFUR. Her current research interests lie in the area of performance analysis of radio resource management techniques in LTE networks and teletraffic of triple play networks. In these areas, she has published several papers in refereed journals and conference proceedings.

E. V. Markova received her B.Sc. and M.Sc. degrees in applied mathematics from the Peoples' Friendship University of Russia in 2009 and 2011, respectively. She is currently a Ph.D. student at PFUR. Since 2012, she works at the Telecommunication Systems Department of PFUR, now she is a Senior Lecturer at the Department of Applied Probability and Informatics of PFUR. Her current research interests lie in the area of performance analysis of radio resource management techniques in LTE networks.

Syllabuses Crawling and Knowledge Extraction of Courses about Global Standardization Education

Hiroshi Nakanishi[1], Tetsuo Oka[2] and Yoshiaki Kanaya[3]

[1]*Osaka University, Japan*
[2]*Tresbind Corporation, Japan*
[3]*Brain Gate Co. LTD., Japan*
E-mail: YFA52730@nifty.com; toka0304@ares.eonet.ne.jp; yoshiaki@kanaya.com

Received December 2014;
Accepted March 2015

Abstract

With progress of social globalization, global standards are becoming more important. This means that more human resources should be developed who are engaged in standardization. To realize lively situation of the global standardization education in universities, it is very important to know the programs, courses and their contents for the education about global standardization in universities. In this paper, firstly, current situation of education about global standardization is studied. As a result, 3 education programs consisting of plural courses and 45 courses were found. Secondary, a new crawling technology to collect syllabuses of courses published on the websites of universities is proposed. Also, a new method of knowledge extraction from the crawled syllabuses is proposed. Using these technologies proposed, syllabuses of 132 Japanese universities including all the 88 national universities were crawled successfully. As a result of the knowledge extraction of the global standardization courses from the syllabuses crawled, it is made clear that 45 courses about global standardization education are offered by 24 Japanese universities. This paper also shows a result of knowledge classification of the 45 syllabuses.

Journal of ICT, Vol. 3, 29–52.
doi: 10.13052/jicts2245-800X.312

Keywords: Standardization, Education program, Crawling, Course, Knowledge.

1 Introduction

Changing ecological system due to global warming, environment contamination by developed industrialization, increase of ecological footprints due to increasing population and decrease of bio capacity make the circumstance of the earth hard.

3 billion of world population in 1960 have increased by 2.3 times to 7 billion in 2011. Human behavior coupled with the activities that need more convenience and more comfort, which has linked with decrease of natural resources and biological products. It has been a big threat to the global sustainability.

To make global society sustainable, it is important to tackle the problem to solve in cooperation with both advanced countries and advancing countries.

One of the important systems to encourage the cooperation of countries is an approach to global standardization.

ICT makes it possible to overcome distance and time between people worldwide, and people can connect to the world while sitting in the comfort of his/her home. Also, ICT contributes significantly to a realization of sustainable global society through the reduction of energy of cars and trains.

Products and services in conformity with the global standards will become available for users everywhere in the world, and global standards can realize the bulk production with stable low price to the world markets and will advance efficient consumption of resources and energies in manufacturing.

Additionally, the formulation of global standards for environmental protection and the control of toxic substances will greatly contribute to the sustainable global society.

As described above, the roles of global standards are extremely important to achieve the sustainable global society. To continue formulating global standards, it is necessary to cultivate 'human resources for global standardization' who will act for the formulation of global standards.

So, strengthening and popularization of the global standardization education are highly important.

In this paper, firstly, situations of global standardization education in universities are surveyed and overviewed.

Secondly, new technologies for syllabuses crawling and knowledge extraction from the crawled syllabuses are proposed. Namely, a crawler collects syllabuses from two different types of syllabus web-pages of universities and extracts a lump of knowledge contained in each of the syllabuses by morphological analysis and unrelated words filtering technologies.

In third, by using a system incorporate with the proposed technologies, syllabuses of 132 Japanese universities including all the 88 national universities were successfully crawled. As a result, it is clarified that 45 courses about global standardization are being offered by 24 Japanese universities.

This paper is written basing on the paper presented at the ITU Kaleidoscope Academic conference 2014 [1].

2 Survey about Standardization Education Situation

2.1 Global Standardization Education in Universities

Through checking websites of 132 Japanese universities including all 88 national universities, it has found that 45 courses concerning global standardization are offered. Also, it has found that Kanazawa Institute of Technology and Osaka University have been offering global standardization education programs [2 – 6] which consist of plural courses concerning global standardization for graduate students. It is summarized as follows.

1. Graduate major program and certificate program for the development of global standardization strategy professionals at Kanazawa Institute of Technology [4].

    Required number of credits for the graduate major program completion is 36, consisting of 10 credits for the 7 specified courses about global standardization, 18 credits for the 9 fundamental courses including intellectual properties and 8 credits for the seminar. Graduate students who have completed the above 36 credits get the master's degree and the certificate of the global standardization strategy professional course. Also, the certificate program named "International Standardization Strategic professional program" is offered to credited auditors, whose completion condition is 10 credits for the 7 specified courses about global standardization same as above.

2. Graduate minor program and certificate program about global standardization at Osaka University [5].

    Required number of credits to complete both of the programs is 8, which should be obtained by studying more than 4 courses among 14 courses specified by the program. The graduate students who have completed 8 credits get the certificate of the graduate minor program of the global standardization in case of the completion of their master's degree program.

Credited auditors of the certificate program get the certificate when they completed 8 credits by studying more than 4 courses among 14 courses specified by the program.

The purpose of the above global standardization education program is to let the graduate students to obtain the knowledge and abilities that are necessary for global standardization activities [6].

The purpose of the above global standardization education program is to let the graduate students to obtain the knowledge and abilities that are necessary for global standardization activities [7].

About foreign universities, a master's degree program is found as in the following.

3. Master's degree program named "Master in standardization, social regulation and sustainable development" at University Geneva [8] was collected.

Required number of credits for the completion of the program is 90 in ECTS, consisting of 15 credits for the 5 courses about global standardization, 42 credits for the 14 courses relating to environmental policy, global health, governance, public policy and international political economy of standards, 15 credits for courses dependent on the student's choice and 18 credits for the internship and the thesis.

The examination results described above are summarized in Figure 1.

2.2 Situation of Education in Industries, Government and Academic Societies in Japan

In addition to universities, education about global standardization is offered by industries, governments and academic societies in Japan.

As global standardization education at industries, companies that approach global standardization set up training and commendation for their approaches to standardization.

As global standardization education at government and foundation, Ministry of Economy, Trade and Industry and Ministry of Internal Affairs and Communications offer delivery lectures by their employees. Japanese Standards Association and Association of Radio Industries and Businesses develop educational materials and offer delivery lectures about global standardization.

University	Program-attribute	Program name	Qualification/Completion-condition
University of Geneva	Master's Degree-Program	Master in Standardization, Social Regulation and Sustainable Development	Master's Degree :90 ECTS credits
Kanazawa Institute of Technology	a. Master's Degree- Program b. Certificate Program for credited auditors	International-Standardization- Strategic Professional Program	a. Master's degree :36 credits b. Certificate :10 credits
Osaka University	a. Graduate Minor Program b. Certificate Program for credited auditors	Global Standardization Program	a. Certificate : 8 credits b. Certificate : 8 credits

(a) Education courses

University	Number of courses	Course contents	Credits/course
University of Geneva	5	Standards, Management system, Risk-management, Strategic- planning, Conformity assesment	3 ECTS
Kanazawa Institute of Technology	7	Shown in Appendix-Table 1	1~2
Osaka University	5	Same as above	2
Waseda University	4	Same as above	2
Tokyo Institute of Technology	4	Same as above	2
Tokyo University	2	Same as above	2
20 Japanese Universities	23	Same as above	1~2

(b) Education courses

Figure 1   Education programs and courses about standardization in universities.

As standardization investigative commission in collaboration with academic societies, universities and government, "Standardization Education Research Committees" are formed under the Institute of Electronics, Information and Communication Engineers and conduct their activities from 2012. Since March 2013, 'network of universities for standardization education' has been studying the education on global standardization.

3 Syllabuses Crawling and Knowledge Analysis of Courses about Global Standardization in Japanese Universities

Almost all the Japanese universities open syllabuses of courses in their homepages. Although research and development have been done about syllabus processing systems [9 – 12], syllabuses crawling and knowledge extraction have not been reported yet.

To automatically collect syllabuses published on the websites of universities and analyze their contents, use of a commercially available crawler software is very efficient. Functions of a crawler are as follows.

a. To collect information on websites according to the direction given by a user.
b. To extract index words contained in the crawled information by morphological analysis.

By using the functions shown above and adding new technologies, syllabuses collection and the knowledge extraction from the crawled syllabuses become possible.

3.1 Survey on How to Access the Syllabus Pages

To successfully crawl syllabuses opened in the websites of universities, it is necessary for the crawler to know how to access the syllabus websites. So, some survey was done about the structure of the syllabus websites by accessing manually the syllabuses of many universities. The result is that there are two types of syllabus web pages was as shown in Figure 2. From Figure 2, following points should be noticed.

1. Syllabus websites consist of syllabuses-table pages and syllabus pages.

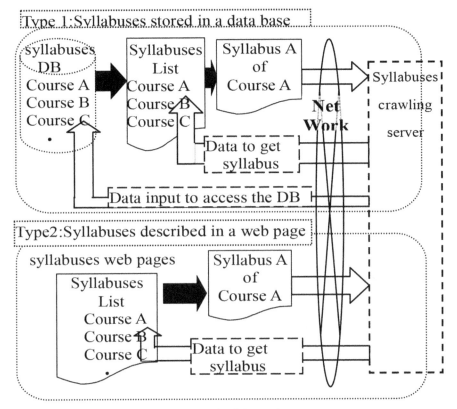

Figure 2   Two types of syllabus webpage structure.

2. Structures of the syllabus websites differ by universities and are classified into two types.

Type 1: Syllabuses are stored in a database and each of them is retrieved by searching the database one by one.

Type 2: Syllabuses are described on the syllabus web-pages and each of them is retrieved by copying the syllabus web-page one by one.

So, crawling requires a procedure guide table (PGT) which gives a crawler the data to access each of the syllabus websites of universities.

In case of crawling type 1 syllabus websites, adequate data input to a database are required twice to get a syllabuses-table and to retrieve syllabuses. In case of crawling type 2 syllabus websites, adequate data input to a web page is required once to get a syllabuses-table.

3.2 Crawling System Design

3.2.1 Crawling method

Using the results described in Section 3.1, crawling methods are designed as in the following.

 1. Crawling method for the syllabuses of Type 1

The first step is to enter data to a database for getting the syllabus-list.

   The second step is to designate one of the syllabuses in the syllabus list and to store it in a storage.

Then, syllabuses are crawled by repeating the second step.

 2. Crawling method for the syllabuses of Type 2.

   As the syllabus list is described in the syllabus web page, syllabuses can be crawled by entering data to the syllabus list and copying each of the syllabuses one by one.

3.2.2 Procedure guide table (PGT) for crawling

The PGT is created by manually accessing the syllabus websites before crawling. and contains data necessary for crawling each of the syllabus websites, such as the first access web address, the type of syllabus webpage, the input data to access to the syllabus pages. The crawler refers to the data in the PGT and crawls syllabus websites.

   University A: First access URL, Type 1, Data 1, Data 2
   University B: First access URL, Type 1, Data 1, Data 2

   University X: First access URL, Type 2, Data

3.2.3 Syllabuses crawling system design

Using the methods described above, syllabuses crawling system is designed as shown in Figure 3.

   In Figure 3, the syllabuses crawling system consists of 4 parts. They are,

 a. Application software for crawling
    It orders the crawler to access the syllabus websites following the procedure guide table data.

 b. Procedure guide table for crawling
    It contains the data for crawling websites of universities

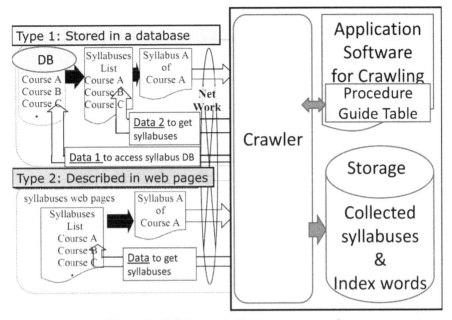

Figure 3 Syllabuses crawling system construction.

c. Crawler

It accesses the web-pages of universities under the order of the application software, collects the syllabuses and stores them to the storage. Also, it extracts index-words from each of the crawled syllabuses.

d. Storage

It is used to store collected syllabuses and their index words generated by the crawler.

Hence, the crawler collects syllabuses automatically and also generates index words contained in the syllabuses by a morphological analysis of the description in the crawled syllabuses. The index words are used for syllabuses searching.

3.3 Design of Knowledge Extraction from Syllabuses

In this section, knowledge extraction method from crawled syllabuses is described.

In a syllabus, much information is described, such as a course name, number of credits, lecturer's name, learning outcomes, weekly lecture plan and various kind of knowledge that a course offers.

Figure 4 shows an algorithm to extract knowledge words from a syllabus.

A crawler automatically extracts index words from each of the crawled syllabuses and store them in the storage. The index words consist of various kinds of words, such as academic knowledge words, lecturer's name, class room and so on.

To extract words about knowledge, it is necessary to remove unrelated words to knowledge from the index words.

For this, a filtering software is introduced which removes unrelated words from the index words

Unrelated words are specified by a user who tries to extract knowledge from syllabuses.

The filtering software extracts knowledge from the index words by removing the unrelated words as shown in Figure 3. In Figure 3, "lecturer's name" and "class room" are specified to be the unrelated words as an example.

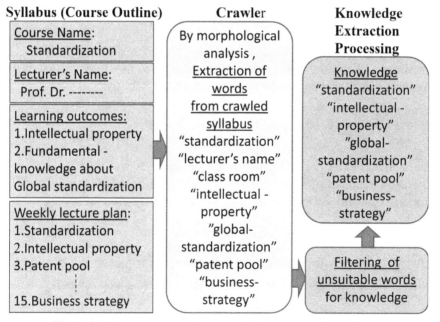

Figure 4   Knowledge extraction algorithm from each of the syllabuses.

3.4 System Design for Syllabuses Crawling and Knowledge Extraction

Using the design explained above, the system for syllabuses crawling and knowledge extraction is designed as shown in Figure 5.

The system consists of 6 parts. They are,

a. Application software for controlling the system
b. Crawler for collecting syllabuses and generating index words
c. Storage for storing the crawled syllabuses, their index words and extracted knowledge words
d. Search engine for searching the index words
e. Filter software for removing unrelated words to knowledge from index words
f. Table of the words unrelated to knowledge

In Figure 5, the crawler collects syllabuses from the syllabus websites of universities and generates index words. The filter software extracts knowledge words from each of the crawled syllabuses and stores them with each of the syllabus.

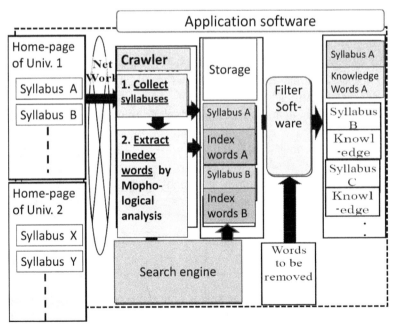

Figure 5 System block-diagram for syllabuses crawling and knowledge extraction.

4 Results of Syllabuses Crawling and Knowledge Analysis

The authors have developed a system named "Interdisciplinary education support system" [13, 14], that implements the design explained in Chapter 3.

Using this "Interdisciplinary education support system", syllabuses crawling was successfully done from 132 Japanese universities including 88 national and 44 major public and private universities.

The reason why crawling targets are only Japanese universities is due to poor information about syllabus websites in foreign universities.

4.1 Selection of Syllabuses about Global Standardization

The crawler collects all syllabuses of universities. So, it is necessary to select syllabuses about global standardization from the crawled syllabuses.

4.1.1 Syllabuses selection related to global standardization

Syllabuses that offer knowledge related to global standardization can be selected by searching the crawled and stored syllabuses using the search engine in the system.

Key words for searching are set by a lump of knowledge required for global standardization. They are,

  a. knowledge concerning "meaning of global standard and related organizations and associations"
  b. knowledge concerning "organizations which formulate global standard and procedure"
  c. knowledge concerning "intellectual properties relating to global standardization"
  d. knowledge concerning "strategies of business and management"
  e. knowledge concerning "national strategies towards global standardization"
  f. knowledge concerning "ability of negotiations"

4.1.2 Result

Syllabuses that contain the knowledge words concerning global standardization were searched among the 132 crawled syllabuses using the keywords of a.~f. shown above.

Table 1 Number of Global standardization courses of universities in Japan

Number of Universities	Number of Courses	
	at Graduate School	at Undergraduate School
24	40	5

As a result, 45 syllabuses of 24 universities were found. 40 syllabuses are offered at graduate schools and 5 at undergraduate schools as shown in Table 1. These 45 syllabuses are still available now.

4.2 Extraction and Analysis of Knowledge about Global Standardization

Knowledge words were extracted from each of the 45 syllabuses by using the method explained in Section 3.4.

Extracted knowledge words are classified into 13 categories of knowledge as shown in Appendix Table 1.

In Appendix Table 1 1, 13 categories of knowledge are written in column wise and 45 syllabuses in line writing direction.

13 categories of knowledge are as follows.

1. Meaning and institutes of global standardization
2. Procedure for formulation of international standards
3. Policy of standardization
4. Human resources for standardization
5. Intellectual properties and patent system
6. Patent pool
7. Management of intellectual properties / strategies
8. Negotiation
9. Communication ability
10. Innovation
11. Research and development /strategies
12. Business model
13. Business competitiveness in international market / strategies / management

Major results are explained below. It should be noticed that the word of 'course' is used instead of the word of 'syllabus' to help readers of this paper imaging the results easier.

From appendix Table 1, it can be understood that each course contains different categories of knowledge. Also, some of the categories of knowledge are common among the courses.

Table 2 shows the number of courses that offer each of the categories of knowledge.

From Table 2 and appendix Table 1, the following points can be understood.

1. The category of knowledge of "Meaning and institutes of global standardization" is commonly offered by 44 courses.
2. The categories of knowledge of "Business model" and 'Intellectual property and patent system' are offered by more than half of the courses.

Table 2   Knowledge classification of courses of universities

Knowledge (Large Classification)	Category of knowledge (Middle Classification)	Number of Courses
Standardization	Meaning and institutes of global standardization	44
	Procedure for the establishment of international standards	14
	Policy of standardization	9
	Human resources for standardization	4
Intellectual properties	Intellectual property and patent system	25
	Patent pool	7
	Management of intellectual properties and strategies	8
Negotiation	Negotiation	9
	Communication ability	6
Research and	Innovation	18
Development	Research and development strategies	12
Business	Business model	31
	Business competitiveness in international market, strategies and management	17

3. The categories of knowledge of "Procedure for the establishment of international standards", "Innovation", "Research and development strategies" and "Business competitiveness in international market, strategies and management" are offered by 30~40 percents of the courses.
4. The category of knowledge of "Negotiation" is offered by 6 courses.

5 Considerations

In this chapter, firstly, new technologies proposed are evaluated. Secondary, the knowledge of courses about global standardization offered by Japanese universities is considered. In third, knowledge of courses about global standardization offered by Japanese universities is considered. In fourth, the difference of study effects between programs and courses is considered.

1. About the technology for syllabuses crawling and knowledge extraction.

In general, web-site crawling is very popular and collects web pages that can be crawled without any special operation such as entering pass-words to websites.

But, when syllabus crawling, it is required to enter proper data to some of the syllabuses websites. The proper data can be obtained by manual access to each of the syllabuses web-sites and are set in the procedure guide table for crawling.

This technology proposed in this paper is new and very essential for syllabuses crawling.

Also, proposed knowledge extraction method from the index words of crawled syllabuses is new and very useful for designing education programs and courses.

2. About the validity of offering courses for global standardization at graduate schools.

In Japanese research universities, almost all the students of science and engineering study at graduate schools. Graduate students have enough times to study some courses for global standardization. Also, graduate students have cultivated abilities to think and understand things such as global standardization with broader perspectives.

So, offering courses for global standardization at graduate schools is reasonable and effective.

3. About the knowledge of courses about global standardization offered by Japanese universities.

To get the knowledge about global standardization, varieties of knowledge are required to study, such as standardization procedures, intellectual properties, research and development, negotiation and business and so on.

As described in Section 4.3, each of the 45 courses about standardization offer some but not all of the 13 categories of knowledge due to the different educational objectives of each of the courses.

It is necessary for the students to select the course from the view point of their study goals.

4. About the difference of study effects between programs and courses.

To understand fully about global standardization, wide range of knowledge are required such as the meaning of standards, intellectual properties, business management, strategies, policies and regulations, negotiation and so on.

Programs are designed to offer wide range of knowledge by the combination of courses. So, studying a program makes it possible for the students to obtain various kind of knowledge systematically.

Single course study is useful for the students to get reviewing knowledge about global standardization.

6 Conclusion

In this paper, firstly, current situation of education about global standardization in universities are surveyed and made clear. Namely,

1. 45 courses about global standardization are offered at 24 Japanese universities.

2. Kanazawa Institute of Technology and Osaka University offer global standardization education programs which consist of plural courses for graduate students and credited auditors.

3. University of Geneva offers a master's degree program of "Master in Standardization, Social regulation and Sustainable Development".

Secondary, new methods for syllabuses crawling and knowledge extraction of courses were proposed. Namely,

4. Survey on the syllabus webpage construction of Japanese universities made it clear that there are two types of syllabus web pages such as database storage type and webpage description type.

5. New crawling technology was proposed that enables to collect syllabuses from two types of syllabus web pages.

6. Also, a new filtering method was proposed that extracts knowledge words by removing unrelated words to knowledge from the index words generated by the crawler.

In the third, syllabuses crawling and knowledge extraction from 132 Japanese universities were executed by using a system that implements the above proposed methods. As a result, following fruitful results were confirmed.

7. Syllabuses were successfully crawled from 132 Japanese universities. Also, 45 syllabuses about global standardization were selected from 132 crawled syllabuses by searching.

8. Knowledge was successfully extracted from 45 syllabuses about standardization and classified into 13 knowledge categories as shown in Table 2 and appendix table 1.

These technologies and results described in this paper will contribute to make the education about global standardization in universities more active and to realize joint education between universities.

One of the further tasks is to crawl syllabuses from universities worldwide.

For this, it is necessary to get syllabus website addresses of universities. They will be gotten from the questionnaire about global standardization education in universities that ITU-T director's Ad.hoc group is now planning [15].

References

[1] Hiroshi Nakanishi, Tetsuo Oka, Yoshiaki Kanaya, "Syllabuses Crawling and Knowledge Extraction of Courses for Global Standardization Education", Proceedings of the 2014 ITU Kaleidoscope Academic Conference, pp.191–196, June 2014.

[2] Hiroshi Nakanishi, "Lump of Knowledge Based Design of Global Standardization Education Program For Graduate students of Universities", The Journal of IIEEJ of Japan, Vol. 42 No 3, pp. 396–400, 2013

[3] Hiroshi Nakanishi, "Graduate Minor education Program on Global Standardization", Proceedings of IEICE General Conference, BP5-2, 2013.

[4] http://www.kanazawa-it.ac.jp/tokyo/ip/ip2.htm

[5] http://www.osaka-u.ac.jp/jp/facilities/gakusai/en/index.html

[6] Hiroshi Nakanishi et al., "Global Standardization Education Program Collaborated by Osaka Univ. and MJIIT UTM", Journal of ICT Standardization, pp. 59–82, Vol 1, 2013.

[7] "Education about Global Standardization in Japan:IEICE Questionnaire Survey", Standards Education AHG – Document 009, Documents 2nd Meeting 20130425-Japan, TSB Director's Ad hoc Group, 2013. http://www.itu.int/en/ITUT/academia/Pages/stdsedu/documents.aspx?RootFolder=%2Fen%2FITU-T%2Facademia%2FDocuments%2Fstdsedu&

[8] http://www.standardization.unige.ch/rationale-for-the-proposed-master-program /course-list.html

[9] Yasuhiko Tsuji, "Information Extraction from Course Syllabi for Automatic Metadata Generation", IEICE Technical Report ET2009-74, 2009.

[10] Fuyuki, Yoshikane, "Syllabus Retrieval Considering Relationship between the Search Term and its Synonyms", Japan Society for Fuzzy Theory and intelligent information Vol. 18, No 2, pp. 299–309, 2006.

[11] Daisuke, Hie, "Development of Kyushu University WEB syllabus cross-searching system", IPSJ SIG Technical Report Vol. 2010-DPS-145, 2010.

[12] Takashi, Kawabata, "Development of General-purpose Syllabus System with Syllabus Object mapping to XML", IPSJ SIG Technical Report Vol. 2009-DPS-141, 2009.

[13] Hiroshi Nakanishi, "Development of the Interdisciplinary Education Support System" , JSET The 26th Annual Conference 2a-508-10, 2009.

[14] Hiroshi Nakanishi et al., "Development of Collecting System of Information on Research and Education of Universities", Journal of the Institute of Image electronics Engineers of Japan, pp. 194–202, vol. 43, no. 2, 2014.

[15] Education about Standardization AHG-Documents 021: Report of the third AHG meeting, June 2014. http://www.itu.int/en/ITU-T/academia/Pages/stdsedu/default.aspx

Appendix

Appendix Table 1. Knowledge extracted from 45 syllabuses about standardization

Univ. (abbrev.)		Osaka					OIT	SIT		NAIST	JAIST	Doshisha	Tokyo		Akita	TMU
Subj. (abbrev.)		Business standard-ization	Intelle-ctual property	Knowledge value society	Topics in Techno-logy	Negotia-tion	Standard-ization and IPR	Global standard-ization	Tech-nology standard-ization	Standard-ization	Tech-nology standard-ization	IPR policy	Innovation and IPR	Standard-ization and ICT	Topics in Standard-ization	Elements of mecha-nism
Included knowledge		Graduate					Graduate	Graduate		Graduate	Graduate	Graduate	Graduate		Graduate	Undergrad
Meaning and Institutes of STD	44	O	O		O	O	O	O	O	O	O	O	O	O	O	O
Proc. for formulation of intl. spec	14	O					O	O	O	O					O	
Standardization policy	9	O														
Human resource quality	4	O														
IPR / patent system	25	O	O				O	O			O	O	O	O		
Patent pool	7	O					O						O			
Management of IPR /strategy	8	O	O										O			
Negotiation	9			O		O								O		
Communication skills	6			O		O										
Innovation	18		O	O	O		O	O					O	O		
Research & development /strategy	12				O	O				O		O	O	O		
Business model	31	O	O	O	O	O			O		O	O	O	O		
Intl. business competitiveness	17	O	O	O	O				O		O		O	O		

Univ. (abbrev.)		KIT								Titech			Tuat	TUS	Akita P.	AIIT
Subj. (abbrev.)		Global standard-ization	Practical tasks for standard-ization	Negotia-tion	Inter-national negotia-tion	Tech-nology standard-ization	Technolo-gy stand-ardization policy	Communi-cation stand-ardization	Standard-ization policy	Standard-ization	Tech-nology standard-ization I	Tech-nology standard-ization II	Industry standard-ization	Standard-ization	Topics in Standard-ization	Standard-ization for IPR
Included knowledge		Graduate								Graduate			Graduate	Graduate	Graduate	Graduate
Meaning and Institutes of STD	44	○	○	○	○	○	○	○	○	○	○	○	○	○	○	○
Proc. for formulation of intl. spec	14	○				○									○	○
Standardization policy	9						○		○					○		
Human resource quality	4		○													
IPR / patent system	25	○	○			○	○			○			○	○		○
Patent pool	7		○			○										
Management of IPR /strategy	8	○	○													○
Negotiation	9			○	○				○							
Communication skills	6	○		○	○								○			
Innovation	18	○				○	○		○	○	○	○				
Research & development /strategy	12							○								
Business model	31	○	○	○	○	○	○		○	○	○	○	○	○		
Intl. business competitiveness	17	○	○										○			○

Univ. (abbrev.)		UEC		Kyushu		Kokushi	Chubu		Chiba	Waseda				K.G.	Yamana	GRIPS
Subj. (abbrev.)		Negotiation for science & tech.	Standardization	Global standardization	Tech. management & standadization	Global standardization	Management for environment	Environmental management	Global standardization	Company and standardization	Communication and standardization	Technology standard	IPR management	Standardization management	Global Standardization	IPR policy
		Graduate		Graduate		Graduate	Undergrad		Undergrad	Graduate			Undergrad	Graduate	Graduate	Graduate
Included knowledge																
Meaning and Institutes of STD	44	○	○	○	○	○	○	○	○		○	○	○	○	○	
Proc. for formulation of intl. spec	14					○	○	○	○	○	○					
Standardization policy	9			○		○				○	○					
Human resource quality	4						○		○							
IPR / patent system	25			○	○	○				○	○	○	○	○		○
Patent pool	7			○	○					○						
Management of IPR /strategy	8										○		○	○		○
Negotiation	9	○										○				
Communication skills	6	○									○	○		○		
Innovation	18	○		○				○				○			○	
Research & development /strategy	12	○	○	○	○							○	○		○	
Business model	31			○	○		○			○	○	○	○	○	○	
Intl. business competitiveness	17			○	○		○	○		○	○	○	○	○		○

Biographies

H. Nakanishi Lecturer, former Professor, Osaka University. He was born in 1947. He graduated from graduate school of engineering of Osaka University in 1973. He received BS and MS degrees from Osaka University and PhD degree from Waseda University. He joined ECL of NTT(Electrical Communication Laboratory of Nippon Telephone and Telegraph public corporation) as a researcher in 1973. His major is electronics and information science. He had been researching and developing Magnetic and Optical storage devices, storage systems and network filing systems. In 2006, he moved to Osaka University as a professor for the interdisciplinary research and education, where he has been researching designs of interdisciplinary education program through analysis of social needs and is teaching a program of Global Standardization. He is a member of The Japan Society of Information and Communication Research, a member of The Institute of Electronics, Information and Communication Engineers, also a member of The Institute of Image Electronics Engineers of Japan.

T. Oka Engineer, former Specially-appointed Professor of Osaka University. He was born in 1949. He graduated from graduate school of engineering of Osaka University in 1973. He received BS and MS degrees from Osaka

University. He joined Mitsubishi Corp. as an engineer in 1973. His major is electronics. He had been researching and developing information systems. In 2006, he moved to Osaka University as a specially-appointed professor for the interdisciplinary research and education, where he had been researching designs of interdisciplinary education systems. He retired from Osaka University in 2012. From 2012, he is working for Tresbind Corp. as a system engineer. He is a member of The Japan Society of Information and Communication Research.

Y. Kanaya Engineer, former Specially-appointed Research Associate of Osaka University. He graduated from graduate school of engineering of Kinki University in 1994. He received BS and MS degrees from Kinki University. He joined Seiyu Corp. as an engineer in 1994. In 2012, he moved to Osaka University as a specially-appointed research associate for the interdisciplinary research and education, where he had been researching designs of interdisciplinary education systems. He retired from Osaka University in 2014. From 2014, he is working for Brain Gate Co. LTD.

Guest Editorial for Special Issue on SDN/NFV Standardization Activities

The telecommunications industry is currently undergoing a radical design change in how to build, manage, and organize networks. Instead of employing the traditional approach of a decentralized management and control plane, a more flexible management paradigm called Software Defined Networking (SDN) is favored. The SDN approach postulates the strong separation of the control plane and data plane of any network element, such as an IP router or layer 2 switch in a data center. Ideally the control plane resides in so-called SDN controllers that control via specialized control protocols the operation of the data plane elements.

This is complemented by transforming networking solutions from a purpose-built box-model design into software-based service functions running on so-called commercial off the shelve (COTS) hardware. This transformation reflects the adoption of IT system design and implementation by the networking industry and is called Network Functions Virtualization (NFV). A network function is composed of a set of function components. All these components are implemented in software for use within a virtual machine. Rapid advancements in virtualization technology, i.e., performance and user acceptance, enabled this movement. The operators will be enabled to on demand instantiate Virtual Network Functions (VNFs). That allows for a significantly reduce time delay of service on-boarding and activation. In addition, VNFs will be able to breath in order to adapt themselves to changing user demands – both my scaling up and down as well as scaling in and out. A flexible service orchestration function together with a lifecycle manager is expected to realize this.

These advancements are not limited to a single technical domain, but spread across the domains of computing, networking, storage, and services in telecommunication networks and beyond. Both domains, SDN and NFV, are seeing a lot of attention in research, product development as well as standardization. Especially, since the transition to this new way of service provision and network operation can be implemented in different evolutionary steps in any of these areas, inter-working with legacy networking elements and physical network functions needs to be foreseen; with that multiple known and new challenges of integrating these new technologies into existing deployments require standardization efforts and are of particular interest.

This special issue provides several viewpoints on these emerging technology shifts by summarizing SDN/NFV standardization efforts from multiple SDOs such ETSI ISG NFV and IRTF.

The first paper of D. Lopez and R. Krishnan presents the recently established Network Function Virtualization Research Group (NFVRG) within the Internet Research Task Force (IRTF). This accounts for the fact that many aspects of SDN and NFV are still in a stage where it is not clear what elements have to be standardized. IRTF Research Groups are tasked to explore technological spaces, which are not yet necessarily ready for standardization activities. The paper describes the process towards the creation of the NVRG, outlines the charter and concludes with an overview about the initial achievements.

K. Jijo George, A. Sivabalan, T . Prabhu, and Anand R. Prasad present and discuss an end-to-end mobile communication testbed, leveraging software that is available as part of various open source projects. The constructed testbed consists of 2G (GSM), 2.5G (GPRS/EDGE) and SAE/LTE network elements, which are operated in a virtualized computing environment. The setting is evaluated against a set of threat scenarios and test cases, and it concludes with the discussions of the results of the scenarios and test cases.

The third paper by B. Wang and M-P. Odini discusses the challenges to enable any type of Virtualized Network Functions (VNF) to work together with an ETSI Network Functions Virtualization (NFV) platform. It proposes a lightweight VNF manager solution for virtual functions in the light of a proof of concept study, based on a selected product of the author's company. A lightweight application, called VNEM (Virtualized Network function Element Manager) is proposed and discussed with the goal to have it open enough to be used by any VNF (Virtualized Network Function). This VNEM provides a VNF independent approach to integrate with any existing VNFs without a new EM (Element Manager) development.

The papers above are providing a snapshot of the current state of the yet evolving fields of SDN and NFV in terms of state of the standardization, early prototyping and proof of concept studies.

However, it is expected that the domains of SDN and NFV are continuing to evolve rapidly, as they conflate the traditionally separated technology domains of telecommunication networks and computing. The challenges ahead are to bridge the different operational understandings of telecommunication networks, e.g., guaranteeing 99,999 % uptime, versus the more relaxed approach of 20 % effort to achieve 80 % of the goal in the computing industry; the mapping of traditional telecommunication functions to virtualized functions in terms of their functionality, performance and reliability; the change in network management to a more comprehensive approach of managing a telecommunication network which is built on top of COTS elements as software artefacts, instead of fixed, specialized telecommunication function elements – amongst many more technological and standardization topics that are still to be addressed.

Martin Stiemerling
University of Applied Sciences Darmstadt
IETF Transport Area Director
martin.stiemerling@h-da.de

Marcus Schöller
University of Applied Sciences Reutlingen
ETSI NFV REL chairman
marcus.schoeller@reutlingen-universiy.de

The NFVRG Network Function Virtualization Research at the IRTF

D. Lopez[1] and R. Krishnan[2]

[1]*Telefónica I+D, Spain*
[2]*Dell, USA*
E-mail: diego.r.lopez@telefonica.com; ramki_krishnan@dell.com

Received 15 April 2015;
Accepted 01 May 2015

Abstract

This paper presents the recently created NFVRG within the IRTF, outlining the story of its constitution, describing and analysing its charter, and presenting the results achieved so far (in despite of its short lifetime) and the intended goals, especially in the short-mid-term.

Keywords: NFV, research, IRTF, IETF, SDN, virtualization.

1 Introduction

The concept of a network essentially implemented in software is a general trend that has consolidated in the recent years and that we can consider based on two main principles:

- Providing general software interfaces for the configuration and usage of network resources thus abstracting the complexity and deployment details of actual network infrastructures.
- Decoupling the different planes conforming the network, and using open interfaces between them, in order to make the supporting infrastructure

Journal of ICT, Vol. 3, 57–66.
doi: 10.13052/jicts2245-800X.313

as much regular and homogeneous as possible, and relying on software mechanisms to support specialized functionalities.

According to these principles, network services are provided by a layered structure, grounded on general-purpose, homogenous hardware, with one or several software layers running on top of it and defining network behavior and functionalities in general. Everything running on the network, from basic functionalities to user applications, and including management and operation elements, becomes a software module that uses the open interfaces provided at each layer.

There are two essential directions that can be applied independently, though they greatly benefit from their simultaneous application, to the degree of being suitable to be considered essentially intertwined. Software Defined Networking (SDN) goes for the decoupling of the control and data planes in a network to gain programmability and simplify data plane elements, while Network Functions Virtualization (NFV) goes beyond, advocating for the general separation of functionality (on software) and capacity (on a general virtualization layer running upon standard regular hardware), increasing network elasticity and drastically reducing the heterogeneity of the supporting infrastructure.

SDN and NFV have been termed as not properly networking technologies, but tools, useful but not at the core of network themselves. While this is arguably true for the application of software to current networking practices, it is also widely acknowledged that they have a strong potential of transforming the way in which networks are designed, deployed and managed. The IETF, focused on the engineering of the Internet through standardization, is exploring the applicability of these technologies as tools to improve different areas of network provisioning and management. However, from the view of the IRTF [1], precisely dedicated to explore the impact of new technologies in the future of Internet, it was natural to address both the long-term implications of these technologies, as well as contribute to consolidate them as essential tools for supporting the general evolution of the network. Given that the concept of SDN predates NFV, the SDNRG [2] was the first created, and the brand new NFVRG followed it.

This paper describes the story of the NFVRG and its consolidation within the IRTF, analyzes in detail its current charter and the connection with other activities related to NFV standardization and development, and details the results achieved so far and the future evolution we envisage for the group.

2 The NFVRG Story

From their initial stages back in 2012, the NFV proponents were well aware of the need of strong research activities and direct collaboration with the research communities to foster a proposal that, while based on the solid ground of current well-established cloud principles, was blazing the trail in a completely new application scenario of these principles. The first NFV whitepaper [3], where the very acronym was coined, lists among the potential benefits of NFV the possibility of a more open and efficient innovation cycle for network infrastructures, where research results could be brought into operational practice in a much shorter time than the current approaches.

The second NFV whitepaper [4], released one year after, directly encouraged "to create new applied research and study programs around NFV". During that year the term NFV had widely spread, and papers, special sessions and workshops started to appear elsewhere. The NFV community formulated a Research Agenda by the beginning of 2014 [5] that included the research topics that were considered most relevant, and that constituted the base for the further NFVRG charter. The first direct references to NFV in the IRTF appeared along the SDNRG meetings at IETF86 in Orlando [6], while the IETF first formally considered NFV during the discussions on the creation of the SFC WG at IETF87 in Berlin [7]. A first introductory session on NFV goals and first steps was held at IETF88 in Vancouver, during a lunch ad-hoc panel discussion. With the increasing of NFV-related research activities, the SDNRG became the natural place for discussing its implications for Internet architecture and protocols at the IRTF, and therefore the number of NFV contributions to the group grew.

The NFV community had started to appoint liaison officers to the most relevant standards organizations in order to foster collaboration, and during IETF89 in London there were several informal discussions among the appointed NFV liaisons to the IETF and the IETF and IRTF leadership. As a result of these discussions, the idea of a potential research group focused on NFV took shape, and an initially small group of people started to sketch the idea, preparing a first proposal for a charter that was shared with the IRTF Chair. After some initial discussion, the IRTF Chair agreed to start NFVRG as a "proposed RG", a tentative status that is used to gauge community interest and potential results of proposed research groups within the IRTF. That allowed the NFVRG proponents to call for a first meeting at IETF90 in Toronto, scheduled out of the agenda along one of the lunch breaks, and a subsequent in-agenda one at IETF91 in Honolulu [8], as well as a much stronger interaction with the IRTF community at large. Two interim meetings were also organized with the main goal of refining the original charter.

The two meetings as proposed RG gather a high participation at all levels: audience, presentations and discussions among the participants. The IETF *datatracker* has listed eight Internet-drafts directly associated with the proposed RG [9] and the charter was gaining in maturity through direct discussion in the NFVRG mailing list. By the end of 2014, the current co-chairs made a direct request to the IRTF Chair to formally charter the NFVRG as an IRTF Research Group. The announcement of the formal charter of NFVRG was made by the end of January 2015.

3 The NFVRG Charter in Detail

The charter of the NFVRG as approved by the IRTF is available at [10]. It discusses the general goals and challenges of NFV research, with special emphasis on its potential impacts on the Internet architecture. As said above, it was originally inspired on the NFV Research Agenda released by the ETSI NFV ISG Technical Steering Committee.

The charter starts with a general declaration of what NFV is and the main characteristics and expected benefits of this technology. Three particular aspects are especially relevant, in defining the main directions for the future work of the NFVRG. First, the possibility of re-thinking current network functions out-of-the-box, building virtual topologies associated with strict functional criteria and fully abstracted from the underlying infrastructure topology, especially in what relates to the constituent nodes. Second, the opportunities for new service provision patterns, where the decoupling between functionality and capacity will enable new roles and new business models. Third is re-thinking the way service provider points-of-presence [13] are managed by exploring opportunities for global optimization across multiple sub-systems beyond networking leading to substantial OPEX savings. While the base technologies (cloud infrastructures and network virtualization essentially) are reasonably established, their application to the provision of network functions is not, and this application domain poses additional challenges that needs to be explored by the research community. The relevance of other research activities within the industry and academia is acknowledged, with a commitment to open collaboration of the NFVRG with technical workshops and scientific conferences, and technology publications.

Besides the consideration of general research problems and specific contributions to standardization activities within IETF WGs, NFVRG will play an important role in contribution of research findings to the standardization efforts in the ETSI NFV ISG, providing exploratory guidance and experimental evidence. Additionally, the NFVRG will help in driving the NFV architectural vision for open source projects in OpenStack [14], OpenDaylight [15], and OPNFV [16] etc. by complementing the NFV standardization efforts in ETSI. Let us consider here and detail the areas of interest included in the NFVRG charter:

- Exploring the new network architectures that can be based on virtualized network functions (VNFs), with a special emphasis in what we could call "network function deconstruction", exploring how to build these functions from virtualized components and identifying compositional patterns.
- How NFV challenges the current cloud architectures, especially in what relates to data plane workloads and the need to adapt VNF deployment to satisfy service topologies and mobility requirements.
- New patterns in network and service function chaining, the architecture implications for chains and the deployment implications for service paths. Extremely relevant issue here are the design and implementation patterns related to virtual and non-virtual functions chaining, and the possibilities of automating it.
- A key aspect of NFV is the orchestration of the virtualization aspects, so the scalability and flexibility associated with cloud-like infrastructures can be achieved. Autonomous orchestration and optimization of VNFs and network services based on them constitute an essential aspect to keep the orchestration problem within manageable limits.
- Reliability and failure characterization is one of the main concerns expressed by many network practitioners when faced with NFV, and thus the NFVRG will study the requirements and mechanisms to ensure reliable virtualized network functions and services.
- NFV will support new operational models, posing a challenge to current practices in network service provider OSS (operation support systems), as well as opening the possibility to streamline them and fostering the application of DevOps [11] paradigm to network operation.
- Virtualized network functions and the infrastructure they run upon will require appropriate abstractions and domain-specific languages to support those levels of abstraction and the openness of the corresponding

interfaces. These include the exploration of such abstractions and the supporting languages and APIs, frameworks for combined processing, network and storage description, policy languages. . .

- It is clear that there will be no NFV zero day, and there is a need to address heterogeneous networks, what in NFV terms implies the coexistence with non-virtualized infrastructure and services, especially when it has to do with management of this heterogeneity and the exploration of the most efficient migration paths.

- Virtualized network functions are expected to play an important role in the convergence of different network technologies and application domains (wired, optical, cellular, satellite. . .) and they have been acknowledged as an essential component of the so-called 5G networks [12].

- Network security concepts are based to a great extent in physical access control (just think about the perimeter concept in current firewalling practice), and this has been brought to the extreme in the case of network infrastructures, where security procedures rely explicitly and even implicitly in the limited physical access to links and nodes. The advent of NFV will shake these assumptions, especially when the network function deconstruction mentioned above is applied. Research in new mechanisms for security, trust and attestation are therefore essential to bring NFV to its full potential.

- Another big hurdle in the NFV path are the claims about inferior performance of VNFs with respect to their physical, traditional counterparts, what NFV proponents reply with the argument that VNFs should only be required to be good-enough for their particular purposes. This implies the need for appropriate performance modeling and characterization.

- Running network functions on a virtualized infrastructure will allow a much more complete and flexible monitoring and data collection (probes can be seamlessly deployed), as well as rely on those collected data to perform adaptive orchestration and management, so real-time big data analytics and data-centric management will be of direct applicability.

- When it comes to this smart management and orchestration, energy efficiency has a specific relevance by its own, especially considered in terms of end-to-end and system-wide optimization of the whole infrastructure (compute, storage and infrastructural network).

- Last but not least, NFV brings the promise of new business models and a deep change in the network service provision economics.

4 Current Results and Future Actions

As we said when outlining the story of NFVRG, in order to focus the activity to maximize impact and address the most pressing research issues, the group has agreed on a few work items to be considered for the near term and therefore fostered by the chairs of the group. This does not imply that contributions addressing any other area in the charter are discouraged, but as an indication of the priorities of the group in the coming months, where results (in the form of research Internet drafts or contributions to other groups) should be produced. These work items are:

1. Policy-based resource management (as depicted in Figure 1 below), addressing optimized resource management and workload distribution based on policy.
2. Analytics for visibility and orchestration, contemplating techniques for the applicability of real-time analytics not only to providing visibility into the NFV infrastructure but also to optimizing resource usage for the purposes of orchestration.
3. Performance modeling focused on facilitating the transition to NFV, so an equivalence model between physical and virtual network functions can be established.
4. Service verification in what relates to security and resiliency, considering aspects related to service and function attestation, and service protection based on elasticity and dynamic management.

Along the two meetings run under the proposed NFVRG, more than twenty presentations were made with lively discussions and a high audience, above 150 attendants. 8 drafts have been submitted, mostly addressing the near-term work items, and several of them are progressing along the IRTF track. Direct links have been established not only with other bodies in the "natural" environment of NFVRG (the IETF and the IRTF as well as the ETSI NFV ISG), but also with open-source projects like OpenStack and the recently launched OPNFV.

As a brief summary, we conclude that we have a huge amount of work ahead with our brand new NFVRG, but it promises to be a more than exciting job. And the recruiting office is always open to those willing to join the (select) club of NFVRGers.

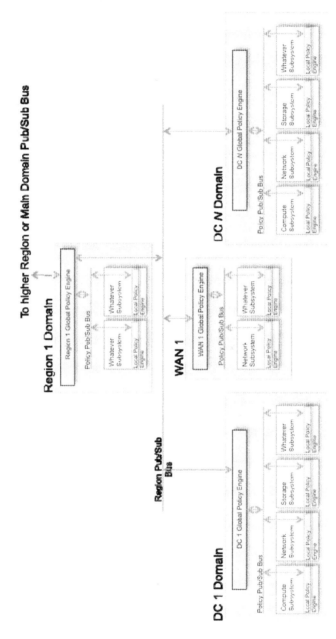

Figure 1   Architecture for policy-based resource management proposed in one of the NFVRG active drafts.

References

[1] "Internet Research Task Force (IRTF)" https://irtf.org/

[2] "Software-Defined Networking Research Group (SDNRG)" https://irtf
.org/sdnrg

[3] "Network Functions Virtualisation – Introductory White Paper" https://
portal.etsi.org/nfv/nfv_white_paper.pdf

[4] "Network Functions Virtualisation – Update White Paper" https://portal
.etsi.org/nfv/nfv_white_paper2.pdf

[5] "ETSI NFV ISG proposed topics for Research Agenda 2014" https:
//portal.etsi.org/Portals/0/TBpages/NFV/Docs/NFV_Research_Agenda_
2014.pdf

[6] "Network Functions Virtualisation" http://www.ietf.org/proceedings/86/
slides/slides-86-sdnrg-1.pdf

[7] "Network Functions Virtualization (NFV)" http://www.ietf.org/proceedin
gs/87/slides/slides-87-nsc-0.pdf

[8] "NFVRG – IETF 91 Honolulu" http://www.ietf.org/proceedings/91/nfvrg
.html

[9] "NFVRG – current IRTF drafts" https://datatracker.ietf.org/rg/nfvrg/docu
ments/

[10] "NFVRG charter," https://irtf.org/nfvrg

[11] Debois, Patrick. "Devops: A Software Revolution in the Making?" *Cutter
IT Journal.*

[12] "NGMN 5G White Paper – Executive Edition by NGMN Alliance"
http://www.ngmn.org/uploads/media/141222_NGMN-Executive_Versio
n_of_the_5G_White_Paper_v1_0_01.pdf

[13] "ETSI NFV Terminology for Main Concepts in NFV" http://docbox.etsi
.org/ISG/NFV/Open/Published/gs_NFV003v010101p%20-%20Termino
logy.pdf

[14] "OpenStack open source project" https://www.openstack.org/

[15] "OPNFV open source project" https://www.opnfv.org/

[16] "OpenDaylight open source project" http://www.opendaylight.org/

Biographies

D. R. Lopez is a Senior Technology Expert on network middleware and services within the GCTO Unit of Telefónica I+D. Diego is currently focused on identifying and evaluating new opportunities in technologies applicable to network infrastructures, and the coordination of national and international collaboration activities. Diego is actively participating in the ETSI ISG on Network Function Virtualization (chairing its Technical Steering Committee), the ONF, and the IETF WGs connected to these activities, acting as co-chair of the NFVRG within the IRTF.

R. Krishnan is an industry expert in the area of Networking and currently Distinguished Engineer and CTO of NFV in Dell. He is a recognized innovator with 20+ US patents and several conference papers. He is co-chair of NFV Research Group in IRTF. He is also a thought leadership speaker in conferences such as ONS, OpenStack, and OpenDaylight etc.

End-to-End Mobile Communication Security Testbed Using Open Source Applications in Virtual Environment

K. Jijo George[1], A. Sivabalan[1], T. Prabhu[1] and Anand R. Prasad[2]

[1]*NEC India Pvt. Ltd. Chennai, India*
[2]*NEC Corporation. Tokyo, Japan*
E-mail: {k.george; sivabalan.arumugam; prabhu.t}@necindia.in;
anand@bq.jp.nec.com

Received 20 January 2015;
Accepted 01 May 2015

Abstract

In this paper we present an end-to-end mobile communication testbed that utilizes various open source projects. The testbed consists of Global System for Mobiles (GSM), General Packet Radio Service (GPRS) and System Architecture Evolution/Long Term Evolution(SAE/LTE) elements implemented on a virtual platform. Our goal is to utilize the testbed to perform security analysis. We used virtualization to get flexibility and scalability in implementation. So as to prove the usability of the testbed, we reported some of the test results in this paper. These tests are mainly related to security. The test results prove that the testbed functions properly.

Keywords: LTE, Amarisoft, Testbed, Security, OpenBTS, OpenBSC, OsmoSGSN, OpenGGSN, OpenIMS.

1 Introduction

Mobile network complexity has increased with time due to the coexistence of multiple technologies like GSM, GPRS, Universal Mobile Telecommunications System (UMTS) and SAE/LTE. Telecom service providers are trying to accommodate the existing customer service (GSM, i.e. voice only) and also

Journal of ICT, Vol. 3, 67–90.
doi: 10.13052/jicts2245-800X.314

trying to provide cutting edge services such as Live TV, video conference etc. without discontinuing the old services. Coexistence of technologies provides more avenues to attack the network by making use of the weaknesses existing in the old generation network. Thus a detailed security analysis of the mobile network is necessary. For this purpose we have developed an end-to-end mobile communication security testbed using open source components on a virtualized platform.

There are a number of projects which have shown the feasibility of using open source components and low cost hardware platform to develop GSM testbed and SAE/LTE testbed. In one such project, an Open Base Transceiver Station (OpenBTS) ported on a common PC with Software Defined Radio (SDR) hardware was used to create a GSM network [1, 2]. Similar to OpenBTS implementation there are other projects which make use of open source for implementing a cellular network that can be operated at low cost, such as private networks in rural deployments, remote areas, or developing countries [3, 4]. During 2011, Anand *et al.* did OpenBTS experiments in the USA based laboratory environment. These experiments accelerated research on OpenBTS for rural deployment, in collaboration with LinkNet/UNZA [4]. In their paper, Anand *et al.* expand on the rationale for OpenBTS, and describe the technical performance that can be expected in a mixed traffic environment, using traffic patterns observed in Macha. Kretchmer *et al.* evaluated the Quality of Service (QoS) of OpenBTS mobile calls across a multi-hop wireless testbed that carries typical Internet traffic [5]. Similar to GSM experimentation there are few projects focused on SAE/LTE technology. One such project is the Amarisoft LTE 100 which is a low-cost SAE/LTE base station running on a Personal Computer(PC) [6], another project is the Open Source Long-Term Evolution Deployment (OSLD) which is a project aimed at design and development of a complete open source SAE/LTE stack [7].

Based on the above literature we see that all of the above projects were either for setting up GSM or SAE/LTE testbeds independently for performance analysis, application trial and QoS experiments. Not been much work has been done on developing an end-to-end mobile communication testbed which houses 2G, 2.5G and SAE/LTE network elements. Thus we have developed an end-to-end mobile communication testbed consisting of 2G, 2.5G and SAE/LTE mobile network elements using open source components on a virtual platform.

The organization of the paper is as follows. In Section 2, we give an overview of the standard 3rd Generation Partnership Project (3GPP) architecture and introduce the concept of virtualization. We present the role of hypervisor in virtualization, the types of hypervisor and how it helps in improving the scalability of our testbed. In Section 3, we discuss the testbed architecture. We mapped the testbed to the standard 3GPP architecture as defined in TS 23.002 [8]. We also explain the open source projects and open components that have been integrated to set up the testbed architecture. In Section 4, we discuss sample threat scenarios and test cases together with the results. Conclusion and future work are discussed in Section 5.

2 Background

The 3GPP architecture has continuously evolved over the years with the changes in the technology. The motivation for the paper is to incorporate the different generations of mobile technology elements on a common testbed to study the security implications. In this section we discuss the key elements of 3GPP architecture and different types of virtual platforms that can be used to create a testbed. Use of virtualization allows us to run different implementation over the same platform without porting the code.

2.1 3GPP Standard Architecture

3GPP scope is to develop and maintain specifications on mobile communications system such as GSM, GPRS, Enhanced Data-rate for GSM Evolution (EDGE), IP Multimedia Subsystem (IMS) and SAE/LTE [9, 10]. Standard architecture of 3GPP is as depicted in the Figure 1. Broadly it constitutes of Core Network (CN) and Access Network (AN) Elements [8]. A brief description of each category is given below:

2.1.1 Core network

The CN elements can be seen as the basic platform for all communication related services provided to the user. The key functionality of the core network elements include switching of calls, routing of packet data, management of user data, authentication and other services. The CN is constituted of a Circuit Switched (CS) domain and a Packet Switched (PS) domain (which includes GPRS and Evolved Packet Core (EPC)). A "CS connection" is a connection for which dedicated network resources are allocated at the connection establishment and released at the connection release. A "PS connection" transports the user information using autonomous concatenation of bits called packets: each packet can be routed independently from the previous one.

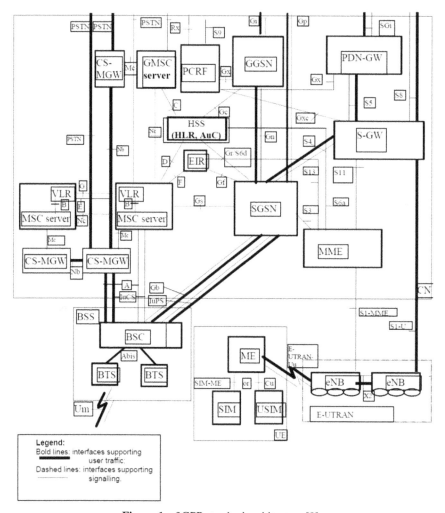

Figure 1 3GPP standard architecture [9].

CN elements common to all technologies GSM, GPRS and SAE/LTE [8]: Home Subscriber Server (HSS): The HSS is the master database which contains the subscription-related information. Function of HSS includes mobility management, call and/or session establishment support, user security information generation, access authorization, service authorization support, etc. There are two subsets of the HSS – Home Location Register (HLR) is a subset of the HSS that enables subscriber access to the CS and PS Domain services and to support roaming to legacy GSM/UMTS CS Domain networks.

Authentication Centre (AuC) is a subset of the HSS that stores an identity key for each mobile subscriber registered with the associated HLR. This key is used to generate security data for each mobile subscriber.

Equipment Identity Register (EIR): An EIR in the GSM system is the logical entity which is responsible for storing the International Mobile Equipment Identities (IMEIs), used in the GSM EDGE Radio Access Network (GERAN)/Universal Terrestrial Radio Access Network (UTRAN)/Evolved Universal Terrestrial Radio Access Network (E-UTRAN) system.

Element specific to PS and CS domains (GSM and GPRS): *Visitor Location Register (VLR)* is the element in a CN which stores the location information of a User Equipment (UE) when it moves out of its home location, the VLR and the HLR exchange information to allow the proper handling of CS calls involving the MS.

Element specific to CS domain (GSM): *Mobile-services Switching Centre (MSC)* is an exchange, which performs all the circuit switching services and signaling functions for mobile stations located in a geographical area designated as the MSC area. MSC acts as the interface between the radio system and the fixed networks.

Elements in PS domain (GPRS): *Gateway GPRS Support Node (GGSN)* is responsible for the inter-networking between the GPRS network and external packet switched networks, like the internet and X.25 networks. GGSN converts incoming data traffic from UE (via the SGSN) and forwards it to the relevant network, and vice versa.

Serving GPRS Support Node (SGSN) is responsible for the delivery of data packets from/to the mobile stations within its geographical service area. Functions of SGSN include packet routing, packet transfer, the mobility management (attach/detach and location management), logical link management, authentication and the charging functions. The location register of the SGSN stores location information (e.g., current cell, current VLR) and user profiles (e.g., International Mobile Subscriber Identity (IMSI), address(es) used in the packet data network) of all GPRS users registered with it; as defined in TS 23.016 [11] and TS 23.060 [12].

Elements in SAE/LTE: *Mobility Management Entity (MME)* – MME is the control plane entity within EPS supporting functions like Non-Access Stratum (NAS) signaling and security, inter CN node signaling for mobility between 3GPP access networks, PDN Gateway (PGW) and Serving Gateway (SGW) selection, roaming, authentication, bearer management functions, etc. MME is responsible for authenticating user towards the HSS. Its duties include authorization of UE to Public Land Mobile Network

(PLMN) and enforcing UE roaming restrictions if any. MME is also the termination point of ciphering and integrity protection for NAS signaling. Lawful Interception (LI) of signaling is also managed and supported by the MME.

Serving Gateway (SGW) – The SGW is the gateway which terminates the interface towards E-UTRAN for user plane. SGW is responsible for data transfer in terms of all packets across user plane. Its duties include taking care of mobility interface to other networks such as 2G/3G.

PDN Gateway (PGW) – The PDN GW is the gateway which terminates the SGi interface towards the PDN. PGW is responsible to act as an "anchor" for mobility between 3GPP and non-3GPP technologies. PGW acts as the point of entry/exit of traffic for the UE. The PGW manages policy enforcement, packet filtratng for users, charging support and LI.

2.1.2 Access Network

The AN elements are the radio interface part of the architecture. Three different types of access network are used by the CN: the AN include GERAN (also called Base Station Subsystem (BSS)), UTRAN (also called Radio Network Controller (RNS)) and E-UTRAN. The MSC can connect to one of the following Access Network type or to both of them: BSS, RNS. The MME and SGW connect to the E-UTRAN.

Access Network Element for GSM we have *Base Station System (BSS)* which is the system of base station equipment (transceivers, controllers, etc.) that is responsible for communicating with UEs in a given area. It constitutes of a Base Station Controller (BSC) with the function to control one or more BTS and a Base Transceiver Station (BTS) is a network element which serves one cell.

Access Network elements for E-UTRAN(SAE/LTE): *E-UTRAN Node B (eNB)* is a logical network element which serves one or more.

E-UTRAN cells: It acts as the radio interface for the SAE/LTE network. An eNB hosts the following functions: Radio Resource Management, Dynamic allocation of resources to UEs in both uplink and downlink, IP header compression and encryption of user data stream, routing of User Plane data towards Serving Gateway. The Evolved UTRAN (E-UTRAN) consists of eNBs, providing the E-UTRA user plane (Packet Data Convergence Protocol (PDCP)/Radio Link Control (RLC)/Medium Access Control (MAC)/Physical Layer (PHY)) and control plane Radio Resource Control (RRC) terminations towards the UE. The eNBs can be interconnected with each other by means of the X2 interface.

2.2 Virtual Environments

Virtualization provides the ability to run multiple operating systems on a single physical system and share the underlying hardware resources. It brings significant cost of ownership and manageability benefits. The virtualization layer sits on top of a software or firmware called hypervisor acting as the intermediary between the physical hardware and the virtual machines (running guest OSs) as shown in Figure 2.

Benefits of virtualization includes: the isolation of virtual machines and the hardware-independence that results from the virtualization process, and also reduce the hardware cost, optimization of workloads, IT flexibility and responsiveness. Virtual machines are highly portable, and can be moved or copied to any industry-standard hardware platform, regardless of the make or model. Thus, virtualization facilitates adaptive IT resource management, and greater responsiveness to changing business conditions [13].

Virtualization, involves a shift in thinking from physical to logical, as it improves resource utilization by treating physical resources as pools from which virtual resources can be dynamically allocated. The following section talks about hypervisor and its role in implementing virtualization.

2.2.1 Hypervisor

Hypervisor is a software or firmware component that can help to virtualize the system resources [14]. The term hypervisor call refers to the

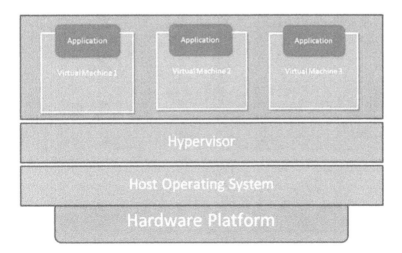

Figure 2 VM architecture.

para-virtualization interface, by which a guest operating system accesses services directly from the higher-level control program. Each operating system appears to have the hosts processor, memory, and resources to itself. In fact, the hypervisor is controlling the host processor and resources, distributing what is needed to each operating system in turn and ensuring that the guest operating systems (virtual machines) are unable to disrupt each other [15].

2.2.2 Hypervisor Classifications

Hypervisor can be classified into two types [13] as shown in Figure 3:

Type 1 hypervisor Bare Metal/Native Hypervisor: Software systems that run directly on the hosts software as a hardware control and guest operating system monitor. A guest operating system thus runs on another level above the hypervisor. This is the classic implementation of virtual machine architectures. Examples of Type 1 hypervisor include VMware ESXi, Citrix XenServer and Microsoft Hyper-V hypervisor.

Type 2 hypervisor Embedded/Host Hypervisor: Software applications that run within a conventional operating system environment. Considering the hypervisor layer being a distinct software layer, guest operating systems thus run at the third level above the hardware. As is done in the case of VMWare's Workstation, Oracle VM Virtualbox.

In this testbed, we used Type 2 hypervisor (i.e. Oracle Virualbox) because the Oracle Virtualbox is a free cross-platform desktop virtualization tool having a host based hypervisor which runs on top of the host operating system (Fedora 17 and Ubuntu 12.04).

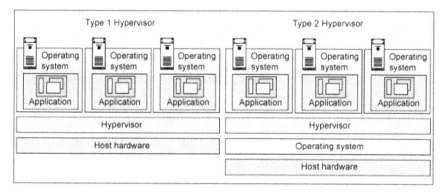

Figure 3 Types of hypervisor.

3 Testbed Architecture and Implementation

In this section we present the testbed architecture and the open source implementations used in the testbed. The details about all of the open source projects are discussed in the subsequent sections.

3.1 Testbed Architecture

Figure 4 shows the overall testbed architecture where we mapped the elements and interfaces of 3GPP standard network architecture, to create an end-to-end testbed in such a way that we can communicate between two different mobile communication technologies.

This testbed has four major group of network elements: i. GSM Network elements, ii. GPRS Network elements, iii. SAE/LTE Network elements, and iv. IMS Network elements.

The GSM network consists of two major network elements, they are: the AN consisting of BTS, and the CN consisting of HSS and MSC as defined in the TS 23.002 [8]. Our testbed architecture makes use of open source projects available for each category of GSM elements which is shown in the Figure 4. We use OpenBTS project which acts as the transceiver and the OpenBSC project which runs in the (Network in the Box) NITB mode it includes the functionality of BSC, MSC, HLR, AuC and HSS. In NITB mode only the Gb interface is exposed for external connection [16, 17].

The GPRS network has two network elements which are implemented in the testbed. In case of GPRS we identified OpenGGSN and OsmoSGSN

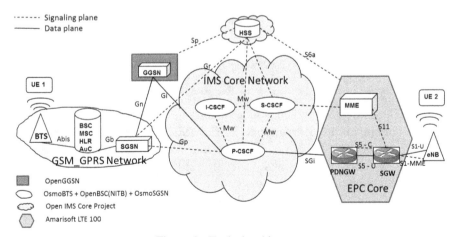

Figure 4 Testbed architecture.

as open source projects to implement the GPRS core network in our testbed. OsmoSGSN is a part of the OpenBSC project under Osmocom and implements an SGSN. OpenGGSN is an open source implementation of GGSN CN element. It connects to OsmoSGSN over the Gn interface and the Gi interface is connected to OpenIMS which acts as the PDN.

The System Architecture Evolution (SAE) is the core network of SAE/LTE. It is an evolved form of its legacy GPRS Core Network. There are no open source projects available for implementing SAE/LTE network elements so we used LTE Amarisoft 100 as the EPC in our testbed.

IMS was developed as an all-IP system designed to assist mobile operators deliver next generation interactive and interoperable services [18]. For our testbed we have used IMS as the core to interlink the two communicating networks. For this purpose we identified OpenIMS as an open source IMS project.

3.2 Open Source Projects and Tools

We have carried out detailed survey of open source projects related to mobile network elements and mapped the relevant projects to the core 3GPP architecture. The various open source and commercial components that we used to setup our testbed providing an end-to-end connectivity between two different mobile communication technologies such as OpenBSC, OsmoSGSN, OpenGGSN, OpenIMS, Amarisoft LTE are given below [16–30].

3.2.1 OpenBSC

OpenBSC is an open source implementation of the BSC features of a GSM network it also includes support for mobility management and authentication and intra-BSC handover, SMS and voice calls. GPRS and EDGE support are possible if combined with OsmoSGSN and OpenGGSN as has been shown in the testbed. It can work as a pure BSC or as a full network in a box.

OpenBSC-NITB mode includes functionality normally performed by the following elements of a GSM network: BSC, MSC, HLR, AuC, VLR, EIR. NITB mode of OpenBSC also implements the A-bis protocol as defined in the GSM TS 08.5× [19–22] and TS 12.21 [23] for communicating with the BTS. It implements a minimal subset of the BSC, MSC and HLR. It does not implement any of the interfaces (like the A and B interfaces) between the higher order GSM network elements.

3.2.2 OsmoSGSN

OsmoSGSN is an open source implementation of the SGSN [24]. It implements the GPRS Mobility Management (GMM) and SM (Session Management). The SGSN connects via the Gb-Interface to the BSS and it connects via the Gn to OpenGGSN. Presently no authentication is done, i.e. the GPRS network will simply allow every IMSI to attach to it as far as it has the same MCC/MNC as the network.

3.2.3 OpenGGSN

OpenGGSN is a Gateway GPRS Support Node (GGSN). The GGSN is a small application which is provided in order to test and demonstrate the use of gtplib. It is fully compliant to the 3GPP standards, but lacks functionality such as charging and management [25]. The project also developed an SGSN emulator suitable GPRS core network testing. OpenGGSN was developed and tested using Redhat 8.0 and 9.0.

3.2.4 Amarisoft LTE

Amarisoft LTE 100 is a LTE Base Station running on a PC, with the configuration mentioned in Section 3.3.2 on the SAE/LTE PC Hardware Requirements. LTEENB, a module of Amarisoft, allows building a real SAE/LTE base station using a standard PC and a low cost software radio front-end. All the physical layer and protocol layer processing is done in real time inside the PC, so no dedicated SAE/LTE hardware is necessary. It can be used to set up an entire SAE/LTE open network.

LTE-MME is a MME implementation [26]. It has a built-in SGW, PGW and HSS. It can be used with the Amarisoft LTE eNodeB.

LTE-ENB is a LTE base station (eNodeB) implemented entirely in software and running on a PC [27]. The PC generates a baseband signal which is sent to a radio front end doing the digital to analog conversion. The reverse is done for the reception. LTE-ENB interfaces with a SAE/LTE Core Network through the standard S1 interface. In particular, the Amarisoft Core Network software (LTEMME) can easily be connected to it to build a highly configurable LTE test network. Some of the implemented features are as follows: implements SAE/LTE release 8 with Frequency Division Duplexing (FDD) configuration, bandwidth ranges from 1.4 to 20 MHz, runs in real time on a standard PC using Linux, Core Network emulation is implemented so that no LTE network infrastructure is needed to use the base station, IP traffic is redirected to a Linux virtual network interface, supports test USIM cards using the standard XOR authentication algorithm, flexible configuration system to support various SAE/LTE parameters. It implements the LTE PHY, MAC, RLC, PDCP, RRC and NAS layers.

3.2.5 OpenIMS

The IMS is an architectural framework for delivering IP multimedia services [28]. The Open Source IMS Core project was developed by the Fraunhofer Institute FOKUS. Its purpose is to provide an IMS core reference implementation for IMS technology testing, IMS application development and prototyping. It's not meant for commercial product development activities. The Open Source IMS Core consists of Call Session Control Functions (CSCFs), the central routing elements for any IMS signaling, and a Home Subscriber Server (HSS) to manage user profiles and associated routing rules [29]. The central components of the Open Source IMS Core project are the Open IMS CSCFs (Proxy, Interrogating, and Serving) which were developed at FOKUS as extensions to the SIP Express Router (SER). As a basic implementation for HSS signaling compatible with SER there is the FOKUS Home Subscriber Server (FHoSS) which is also part of the Open Source IMS Core project [30].

3.3 Testbed Implementation

Figure 5 shows the implementation view of the end-to-end testbed discussed in the previous section. This implementation mainly consists of two generations of mobile network elements i.e. 2G/2.5G and SAE/LTE. We used mainly open source projects, operating systems, open source hypervisor and low cost open source hardware platforms.

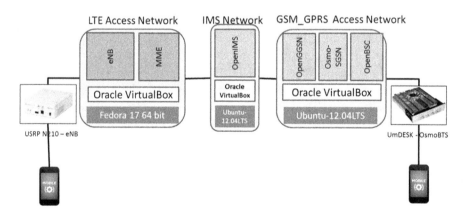

Figure 5   End-to-end testbed architecture.

The GSM-GPRS implementation involves a small base transceiver station utilizing OpenBTS, OpenBSC, OpenGGSN and OsmoSGSN software [31]. The Linux- based software application configures the UmTRAY to present a GSM air interface to standard GSM phones and also interfaces with the OpenIMS on the back-end acting as PDN.

For the SAE/LTE implementation we configured the Universal Software Radio Peripheral (USRP) hardware using Linux based USRP Hardware Drivers (UHD). We ran the LTE-MME and LTE-ENB on the PC and the radio signal trans-reception carried out on the USRP. Similar to GSM-GPRS we connected the SAE/LTE core network to the same OpenIMS network thus enabling the end-to-end connectivity between GSM-GPRS and SAE/LTE network.

We do not have the spectrum license to perform experiments in the licensed spectrum, so we directly connected the UE to the BS hardware via RF cables, which ensured that there is no RF radiation due to this experiment. The following section explains in detail about the hardware and software used in this implementation.

3.3.1 GSM/GPRS Testbed
We set up the OpenBSC-GPRS testbed using a number of open source hardware and software elements, as given below:

Hardware
- UmTRAY as an SDR to act as GSM BTS.
- A Fast PC: Ubuntu 12.04 LTS OS, 1 Gigabit Ethernet ports, 2 GB of RAM, 1 GB of hard disk space.

Software
- Oracle Virtualbox 4.3.12 as hypervisor.
- Open source projects on Base Station Controller (OpenBSC).
- Open Source BTS software OsmoBTS.
- Open source transceiver OsmoTRX for GSM Layer 1 implementation.
- Open source packet control unit (OsmoPCU) RLC and MAC layers of the GPRS Um (radio) interface on the MS-facing side, as well as the Gb Interface (NS, BSSGP) on the SGSN-side.
- OpenGGSN an Open Source GGSN Implementation.
- Off the shelf mobile phones with GSM SIM cards as client devices.

3.3.2 LTE Testbed

We set up the SAE/LTE Access network using the Amarisoft LTE 100, other hardware and software, as mentioned below:

Hardware
- A Fast PC
    - A quad core Intel Core i7 CPU (Nehalem or later).
    - 2 Gigabit Ethernet ports.
    - 2 GB of RAM.
    - 1 GB of hard disk space.
- Radio front end.
    - USRP N200 or N210 from Ettus Research with the SBX daughterboard[32].
    - Antennas for the intended LTE frequencies or cables and attenuators to connect to a UE.
- LTE UE compatible with LTE release 8 FDD.
- Test USIM cards.

Software
- A 64 bit Linux Fedora 17 [33].
- Oracle Virtualbox 4.3.12 as hypervisor.
- UHD drivers ($\times 86$ 64 target) from Ettus Research [28].
- The Amarisoft LTE Core Network.

4 Sample Tests and Results

To showcase the usability of this testbed (i.e. security testing), in this section we discuss sample threat scenarios and test cases together with the results. The test cases, given in following sub-sections, are only examples to prove the usability of the testbed, other test cases can be tried as well. The definitions of some of the terminology used in the test cases are given below.

Theft of service: In this form of attack a malicious user gains access to a legitimate user's services. The attacker can then use these services while the user pays for them without realizing it being misused.

Information Disclosure Vulnerability: Information Disclosure is a loophole in a network where the information may be leaked outside the network. This can lead to crucial information being made available to an attacker.

Integrity Check: Integrity is one of the basic principles of security, which deals with maintaining the accuracy and consistency of the data being communicated. The loss of integrity can be caused by someone gaining access to the information in transit wherein he can modify the data and forward it without the end users realizing it.

4.1 Test Case I – Unauthorized Access and Theft of Service

- *Scenario:* The UE is assigned an IP address dynamically by the MME from an IP address range when setting up the Packet Data Protocol (PDP) context. The UE then accesses the services using the given IP address. In this test case we manually assigned an IP in the same subnet to our network interface and used it to communicate with the PGW.
- *Expected Result:* The UE must not be able to modify the assigned IP address, as this may lead to users being able to steal other users' services theft of service and/or unauthorized access. The UE should only be able to communicate via its assigned IP address.
- *Observed Result:* In the given test case we statically assigned an IP in the valid range to the UE and use it for communicating with the MME as shown in Figure 6 and 7. We saw that the IP address is assigned as a virtual IP in Windows and pinging via the static IP to the core network was unsuccessful. Here 192.168.3.2 was our assigned IP while 192.168.3.20 was the static IP allotted by us. We observed that communication with the core network was only possible when using the network

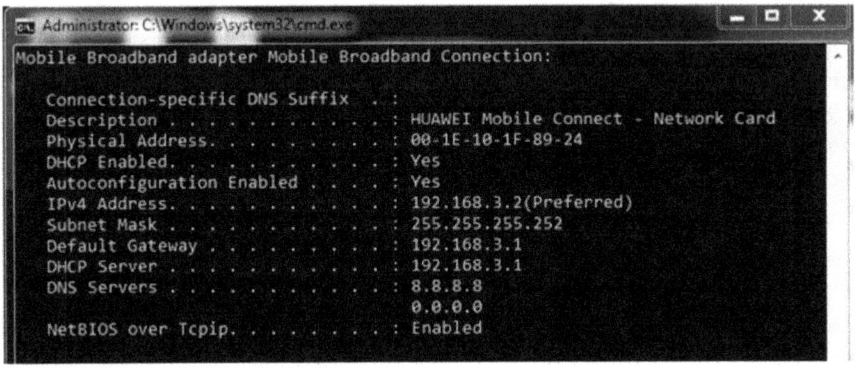

Figure 6   The default IP configuration.

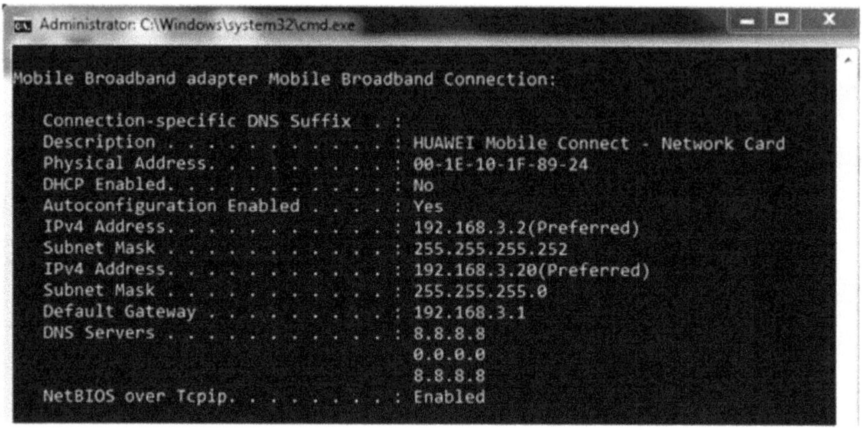

Figure 7   The manual IP assignment testcase.

assigned IP i.e. 192.168.3.2, thus proving that the implementation is secure.

4.2 Test Case II – Information Disclosure Vulnerability

- *Scenario:* Scanning the IP addresses using any network sniffer we can sniff for the communication between the UE and the PGW. As a simple test on information disclosure vulnerability we check whether network information, in form of IP address, is sent to the UE.
- *Expected Result:* In a secure mobile communication system there should be no leakage of information on network topology. The communications between the UE and PGW should not contain any such protected information about the core network.
- *Observed Result:* We captured the packets in the UE interface using wireshark as shown in Figure 8. Based on the observation of the captured pcaps file; data from the protocol hierarchy, contents summary and endpoints we found information about one of the network IP interfaces i.e. 192.168.3.3 apart from our tunnel IP range. As we already know that our IP address is 192.168.3.2 from the previous test case, the detected IP address is an interface to the core network. This core network IP should not have been visible to the UE and leads to Information Disclosure vulnerability.

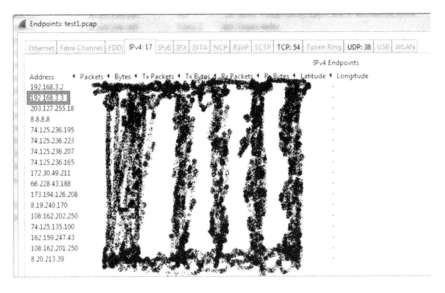

Figure 8   The core network IP disclosure.

4.3 Test Case III – Integrity Check

- *Scenario:* To ensure integrity, the receiver verifies that the received NAS message is exactly the message that the transmitter sent. This is done using an integrity value which is derived using the KNASint Key. MME initiates the NAS security procedure by sending the message which includes encryption and integrity protection algorithms. Key selection identifier (KSI-ASME) is also included in the message. UE responds back to the MME with a message which is ciphered and integrity protected.
- *Expected Result:* The SAE/LTE architecture mandates that the NAS signaling between the UE and MME be integrity protected. Apart from the first communication of the UE, the IMSI is not sent over the network only the Globally Unique Temporary Identifier (GUTI) is used for communicating, and setting up of the encryption and integrity keys between the UE and the MME.
- *Observed Result:* In our testbed on scanning the MME and ENB logs we were able to monitor the NAS signaling data gaining information about the transmitted attach request for setting up the initial connection request.

The NAS message as given in Figure 9 is integrity protected to ensure the origin of the sender but not ciphered to protect the communication. It uses the GUTI to set up the attach process with the MME which can be seen in the NAS messages.

The Figure 10 shows NAS signaling Attach accept setup once the integrity and ciphering has been setup. It shows a secure communication being setup with integrity protection and ciphering to protect the signaling information exchange. The NAS message shows the signals sent when setting up user context and exchanging APN info and IP info with the new user.

Figure 9   Unencrypted attach request using GUTI.

```
14:47:17.174 [NAS] DL 0064 EMM: Attach accept

          Protocol discriminator = 0x7 (EPS Mobility Management)
          Security header = 0x2 (Integrity protected and ciphered)
          Auth code = 0x2164dc69
          Sequence number = 0x01
          Protocol discriminator = 0x7 (EPS Mobility Management)
          Security header = 0x0 (Plain NAS message, not security protected)
          Message type = 0x42 (Attach accept)
          EPS attach result = 2 (combined EPS/IMSI attach)
          T3412 value:
            value = 0
            unit = 7 (deactivated)
          TAI list:
            Length = 6
            Data = 00 00 f1 10 00 01
          ESM message container:
            Protocol discriminator = 0x2 (EPS Session Management)
            EPS bearer identity = 5
            Procedure transaction identity = 1
            Message type = 0xc1 (Activate default EPS bearer context request)
            EPS Qos:
              Length = 1
              Data = 09
            Access point name = "test123"
            PDN address:
              PDN type = 1 (IPv4)
              IPv4 = 192.168.3.2
            Protocol configuration options:
              ext = 1
              configuration protocol = 0
              Protocol ID = 0x8021 (IPCP)
              Data = 03 00 00 0a 81 06 08 08 08 08
          GUTI:
            MCC = 001
            MNC = 01
            MME Group ID = 32769
            MME Code = 1
            M-TMSI = 0x00000002
          Location area identification:
            MCC = 001
            MNC = 01
            LAC = 0x0001
          MS identity:
            Length = 5
            Data = 04 00 00 00 02
```

Figure 10 Integrity protected and ciphered attach accept containing the APN and assigned IP.

5 Conclusion and Future Work

5.1 Conclusion

This paper demonstrates creation of an end-to-end testbed for GSM, GPRS and SAE/LTE using open source projects. We implemented the various network elements in a virtual platform to make it scalable and flexible to deploy. In this paper we demonstrated the usability of the testbed based on a few test cases. Some of the sample categories of test cases that have been performed include: theft of service, information disclosure vulnerability and integrity check. Finally, the observed results were compared with the expected results to ascertain any defects in the system. The utilization of this testbed is not only limited to perform security tests, but can also be used for various other studies as well.

5.2 Future Work

Moving forward we plan on taking our whole implementation to a more open architecture implementation using Network Function Virtualization (NFV). As a first step, the virtualization aspects of Linux platform will be implemented and managed using OpenStack. OpenStack is an open source implementation that is used to control large pools of processing, networking and storage resources [34]. The whole system can be managed and resources can be assigned by an administrative dashboard and web interface. This will facilitate us to easily build and manage mobile network elements on a flexible virtual platform.

Further, we plan to implement project Clearwater. This project is specially designed to be scalable over the cloud network to provide IMS solutions to the users. And it does so by using the concept of NFV as it has been built from the ground up to run in a virtualized environment and take full advantage of the flexibility of the cloud [35].

References

[1] Burgess, D. A, Samra H. The Open BTS Project an opensource GSM base station, Sept 2008.

[2] OpenBTS: http://openbts.org/

[3] Heimerl, K., Brewer, E. The village base station. "In Proceedings of the 4th ACM Workshop on Networked Systems for Developing Regions Systems." San Francisco (CA), 2010, pp. 5–6.

[4] Anand A., Johnson, D. L., Belding, E. M. "Village Cell: cost effective cellular connectivity in rural areas. In Proceedings of the International Conference on Information and Communication Technologies and Development". Atlanta (GA), 2012, pp. 180–189.

[5] Mathias Kretschmer, Peter Hasse, Christian Niephaus, Thorsten Horstmann, and Karl Jonas. Connecting Mobile Phones via Carrier-Grade Meshed Wireless Back-Haul Networks. E-Infrastructures and E-Services on Developing Countries. Africomm 2010, 2010.

[6] Amarisoft. Amari LTE 100, Software LTE base station on PC. Available at: http://www.amarisoft.com/

[7] OSLD Project. Open Source Long-Term Evolution (LTE) deployment. Available at: https://sites.google.com/site/osldproject/

[8] 3GPP TS 23.002 – Network Architecture.

[9] Gottfried Punz, Evolution of 3G Networks: The Concept, Architecture and Realization of Mobile Networks Beyond UMTS, Springer Wien-New York, 2010.

[10] Heikkei Kaarannen, UMTS Networks: Architecture, Mobility and Services, John Wiley and Sons, 2005.

[11] 3GPP TS 23.016: "Subscriber data management; Stage 2".

[12] 3GPP TS 23.060: "General Packet Radio Service (GPRS); Service description; Stage 2".

[13] Virtualization – http://www.ibm.com/developerworks/cloud/library/cl-hypervisorcompare/

[14] Hypervisor – http://www.tricerat.com/resources/topics-library/hypervisor-virtualization-software

[15] The term Hypervisor – Gerald J. Popek and Robert P. Goldberg (1974). "Formal Requirements for Virtualizable Third Generation Architectures". Communications of the ACM 17

[16] OpenBTS: http://wush.net/trac/rangepublic

[17] OpenBSC Network from scratch: http://openbsc.osmocom.org/trac/wiki/network from scratch

[18] 3GPP-IMS: http://www.3gpp.org/technologies/keywords-acronyms/109-ims

[19] TS GSM 08.52 BSC-BTS Interface Principles.

[20] TS GSM 08.54 BSC-BTS Layer 1 Specification.

[21] TS GSM 08.56 BSC-BTS Layer 2 Specification.

[22] TS GSM 08.58 BSC-BTS Layer 3 Specification.

[23] TS GSM 12.21 BSC-BTS Operation/Maintenance Signalling.

[24] Osmo-SGSN_OpenBSC: http://openbsc.osmocom.org/trac/wiki/osmo-sgsn

[25] OpenGGSN Readme – http://cgit.osmocom.org/openggsn/tree/README

[26] LTE – MME Document by Amarisoft.

[27] LTE – ENB Document by Amarisoft.

[28] Documentation-OpenIMS.org: http://www.fokus.fraunhofer.de/en/fokus_testbeds/open_ims_playground/components/osims/index.html

[29] Mohammad Ilyas, Syed A. Ahson, IP Multimedia Subsystem (IMS) Handbook, CRC Press, 2009.

[30] Dragos Vingarzan, Peter Weik and Thomas Magedanz, Design and Implementation of an Open IMS Core, Springer Berlin Heidelberg, 2005.

[31] OpenBSC-GPRS Implementation: http://openbsc.osmocom.org/trac/wiki/OpenBSCGPRS

[32] USRP N210 – https://www.ettus.com/product/details/UN210–KIT
[33] Fedora Project Downloads-http://fedoraproject.org/en/get-fedora-all
[34] OpenStack Open Source Cloud Computing Software: https://www .openstack.org/software
[35] Project Clearwater: http://www.projectclearwater.org/about-clearwater/

Biographies

K. J. George received B.Tech in Computer Science and Engineering from Kurukshetra Institute of Technology and Management, India in 2011. He has 2 years of experience in research and development of mobile communication networks. At present he works as Member Technical Staff in NEC India Standardization (NIS) Team at NEC Mobile Network Excellence Center (NMEC), NEC India Pvt Ltd, Chennai. Prior to joining NECI he was as sociated with IIIT, Bangalore as Research Associate. In his current role, he is working on security aspects of telecom networks and testbed development of next generation mobile networks. His research interest includes Penetration Testing, Network Security and Telecom Security.

S. Arumugam received Ph.D in Electrical Engineering from Indian Institute of Technology Kanpur, India in 2008 and M.Tech degree from

Pondicherry University, India, in 2000. He has 14 years of experience in Academic teaching and Research. Presently he works as Manager for Research at NEC Mobile Network Excellence Center (NMEC), NEC India Pvt Ltd, Chennai. Prior joining NECI he was associated with ABB Global Services and Industries Limited, Bangalore as Associate Scientist. He has published more than 25 papers in various International Journals and Conferences and also participated in many National and International Conferences. In his current role, he is representing NEC for Global ICT Standards forum of India (GISFI). His research interest includes Next Generation Wireless Networks.

P. Thiruvasagam received Master of Design in Communication Systems from Indian Institute of Information Technology Design and Manufacturing-Kancheepuram, Chennai, India in 2014. He has 2 years of experience in academic teaching. At present he works as Research Engineer in NEC India Standardization (NIS) Team at NEC Mobile Network Excellence Center (NMEC), NEC India Pvt Ltd, Chennai. His research interest includes Information Security, Wireless Telecom Security and Next Generation Wireless Networks.

A. R. Prasad, Dr. & ir., Delft University of Technology, The Netherlands, is Chief Advanced Technologist, Executive Specialist, at NEC Corporation,

Japan, where he leads the mobile communications security activity. Anand is the chairman of 3GPP SA3 (mobile communications security standardization group), a member of the governing body of Global ICT Standardisation Forum for India (GISFI), founder chairman of the Security & Privacy working group and a governing council member of Telecom Standards Development Society, India. He was chairman of the Green ICT working group of GISFI. Before joining NEC, Anand led the network security team in DoCoMo Euro-Labs, Munich, Germany, as a manager. He started his career at Uniden Corporation, Tokyo, Japan, as a researcher developing embedded solutions, such as medium access control (MAC) and automatic repeat request (ARQ) schemes for wireless local area network (WLAN) product, and as project leader of the software modem team. Subsequently, he was a systems architect (as distinguished member of technical staff) for IEEE 802.11 based WLANs (WaveLAN and ORiNOCO) in Lucent Technologies, Nieuwegein, The Netherlands, during which period he was also a voting member of IEEE 802.11. After Lucent, Anand joined Genista Corporation, Tokyo, Japan, as a technical director with focus on perceptual QoS. Anand has provided business and technical consultancy to start-ups, started an offshore development center based on his concept of cost effective outsourcing models and is involved in business development.

Anand has applied for over 50 patents, has published 6 books and authored over 50 peer reviewed papers in international journals and conferences. His latest book is on "Security in Next Generation Mobile Networks: SAE/LTE and WiMAX", published by River Publishers, August 2011. He is a series editor for standardization book series and editor-in-chief of the Journal of ICT Standardisation published by River Publishers, an Associate Editor of IEEK (Institute of Electronics Engineers of Korea) Transactions on Smart Processing & Computing (SPC), advisor to Journal of Cyber Security and Mobility, and chair/committee member of several international activities.

He is a recipient of the 2014 ITU-AJ "Encouragement Award: ICT Accomplishment Field" and the 2012 (ISC)² Asia Pacific Information Security Leadership Achievements (ISLA) Award as a Senior Information Security Professional. Anand is Certified Information Systems Security Professional (CISSP), Fellow IETE and Senior Member IEEE and a NEC Certified Professional (NCP).

Lightweight VNF Manager Solution for Virtual Functions

Wang Bo[1] and Odini Marie-Paule[2]

[1]ES CMS. Hewlett Packard, Shanghai, China
[2]ES CMS. Hewlett Packard, Grenoble, France
E-mail: {bow; marie-paule.odini}@hp.com

Received 09 October 2014;
Accepted 01 May 2015

Abstract

This paper presents a creative lightweight solution to package any existing VNF (Virtualized Network Function) to support ETSI NFV (Network Function Virtualization) architecture quickly and easily. The preconditions of existing VNF are quite straightforward and simple, such as running on Virtual Machine (VM) and Operating System (OS) supporting Secure Shell (SSH) commands. With our solution, the new VNF (Virtualized Network Function) will have the additional functions required to integrate with ETSI NFV architecture, without redesigning the existing Element Manager. This will include VNF Descriptor, lifecycle management benefits such as deploy within minutes, scale out/in on demand, etc.

Keywords: NFV, VNF, VNFM, EM.

1 Problem Statement

Networks Functions Virtualization (NFV) is a powerful emerging technique with widespread applicability. It is about virtualizing network functions, leveraging general purpose platforms (Compute, Network, and Storage) and virtualization to provide agility, flexibility and decouple network functions from physical resources. It enables CSPs (Communication Service Providers)

Journal of ICT, Vol. 3, 91–104.
doi: 10.13052/jicts2245-800X.315

to radically take down infrastructure capital and operational costs – and provide a more agile environment to introduce new services more quickly at far less cost than previous network infrastructure. The main business benefits includes Cost/ROI, Business agility, Scalability, Flexibility and Innovation. ETSI (European Telecommunications Standard Institute) is working on NFV and has released the first phase specifications [1].

To implement NFV, a Network Function (NF), which is today often a network node or physical appliance and its Element Manager (EM, a component to manage the NF), has to be virtualized as a VNF (Virtualized Network Function). NFV decouples software implementations of Network Functions from the compute, storage, and networking resources they are using. It adds new capabilities to communications networks and requires a new set of management and orchestration functions to be added to the current model of operations, administration, maintenance and provisioning. It requires VNF to provide not only an EM (Element Manager) to take care of the configuration and management of the network functions, but also a VNF Manager. The VNF manager typically takes care of the lifecycle of the VNF. It is tightly coupled to the service logic of the VNF and this intelligence typically comes from the VNF vendor.

HP has multiple telecommunication products, which acts as NFV NF (Network Function). Most of these products are evolving to NFV, and support at least HP NFV Level 2, which means run on a VM. In parallel, HP has a product, NFV Director (NFV-D), for NFV Orchestrator, that embeds a default generic VNF Manager. But no solution exist for VNF level 2 to be sold to customers that have an existing NFV Orchestrator and only want HP VNF and that VNF EM+VNF Manager. Two options have been considered:

First, evolving existing EM (Element Manager) or having VNF team develop a VNF Manager for its product. But this solution does not seem to make sense: it would mean a duplication of effort for each VNF team, time consuming and costly; Some NF have rich management features, but evolving this to NFV may be a challenge for the teams: not all of these features are required for NFV, and it may be difficult to modify the EM code; Other NFs have a lack of management features, and just have some basic capability like CLI. In this case a complete new design is required with often little experience in that team to develop management platforms.

Second, use the NFV-D and its embedded VNF Manager for each product line NF. But this solution was not available when we started the POC (Proof of Concept), and in any case would only work if the customer is ready to buy an integrated solution from HP, with VNF and NFV orchestrator/VNF manager. In case another vendor NFV Orchestrator has been deployed in the CSPs' NFV infrastructure. HP VNF will have to integrate with that NFV Orchestrator.

Besides, today there is not a standard and comprehensive VNF Descriptor to describe a VNF in terms of its deployment and operational behavior. ETSI has defined some specifications for descriptors but it remains guidelines not detailed specifications ready for implementation and it is primarily focused on VNF on-boarding and managing the lifecycle of a VNF instance.

In the meantime, CSPs are moving full speed on NFV, with expectations of reducing cost and increasing agility quickly and expect a lightweight solution to demonstrate some VNF's capability in their existing NFV infrastructure.

2 Our Solution

Our proposed solution is to provide a lightweight application, which is called VNEM (Virtualized Network function Element Manager) that would be open enough to be used by any VNF (Virtualized Network Function). It provides a VNF independent approach to integrate with any existing VNFs without a new EM (Element Manager) development, and also provide a VNF Manager functionality to support de facto standards Openstack with the NFV infrastructure. Figure 1 illustrates the VNEM position in ETSI NFV architecture.

The EM (Element Manager) features, which is required in NFV MANO (Management and Orchestration) architecture, are packaged to VNEM as some artifacts, include some XML based workflow configurations, shell script template, cloud-init template, etc.

For the purpose of this POC (Proof of Concept), we have implemented VNEM for the virtualization of an HP Product called HP OCS (Online Charging System). Figure 2 illustrates the functional architecture of VNEM. It is composed of:

- A STF (SSH and Template based Flow engine), which provides the easy integration with any VNF which has SSH capability; all of the VNF context, such as VNF Descriptor (VNFD), VNF Record (VM address, network ...) information, are available for the template, and rendered before SSH interaction with VNF Components. This approach is widely supported by any OS, which enables our solution to integrate with any existing Network Function easily.

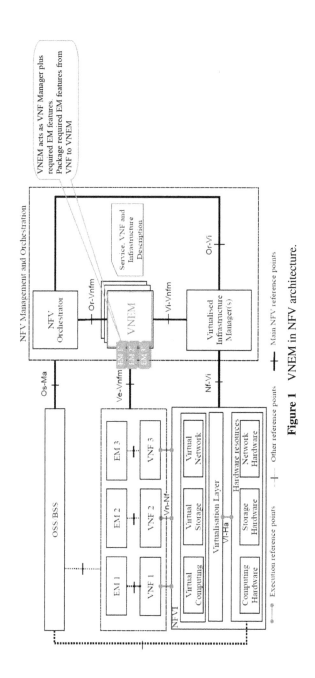

Figure 1   VNEM in NFV architecture.

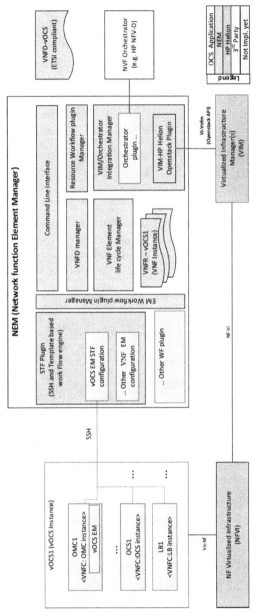

Figure 2 VNEM function architecture for vOCS.

- An Openstack integration plugin with VIM (Virtualized Infrastructure Manager). The VIM in this POC is HP Helion [2], and the Openstack API is de facto standards for VIM integration.
- A resource workflow plugin to compose the VIM interaction tasks. The task types include remote shell, remote copy, and openstack operations.
- A comprehensive and ETSI NFV compliant VNFD and its management. The VNFD describes all required compute, network, storage resources, scalability and operating workflows, and also supports different types of deployments, e.g. small and big deployments.

HP vOCS (Virtualized Online Charging System) is a typical VNF (Virtualized Network Function). Figure 3 illustrates the VNFD (VNF Descriptor) with some key elements we are using to describe HP vOCS. HP vOCS includes 3 VNF Components, LB (Load Balancer), OCS (Online Charging Server) and OMC (Operation, Maintenance and Configuration). Each VNF Components can be instantiated to multiple VM instances based on the defined instance count in VNFD. For the VNF Component which can be scaled out/in on demand, its scalability is defined in VNFD. Each VNF Component has multiple network interfaces and connect to virtualized networks, virtualized networks are defined in VNFD too, and they will be created while the vOCS is created (instantiate). Figure 6 illustrates the HP Helion Dashboard to show a new vOCS is created from scratch based on a VNF Descriptor with small deployment flavor through an instantiate operation.

While the vOCS is created (instantiated), VNEM will install and configure each VM with predefined artifacts, and then start the application on the VM. Those operations are mapped to lifecycle operations.

The VNEM can manage a full lifecycle of a VNF from VNF on-board, instantiate, configure, scale in/out, start/stop, terminate and VNFD demission. Figure 4 illustrates the VNF lifecycle managed by VNEM.

Figure 6 also illustrates HP Helion Dashboard with vOCS operations (Instantiate, Scale in/out, Terminate), e.g. the scale out operation to create a new VM. The scale out operation can be triggered by VNEM, or by an external NFV Orchestrator, it can be automatic based on the runtime data of vOCS, e.g. the traffic throughputs, or it can be manually triggered.

VNEM provides a CLI (Command Line Interface) to operate on the VNF and its VNF instances.

One VNEM can manage multiple VNF and its multiple VNF instances.

Figure 3 VNFD (VNF Descriptor) Example for vOCS (key elements only).

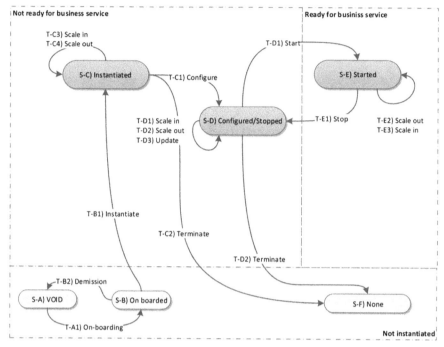

Figure 4   VNF lifecycle with VNEM.

Once a VNF is instantiated, a VNF instance will be created based on the specified VNFD and deployment flavor, and a new VNFR (VNF Record) will be created. The VNFR includes the logical and physical resources allocated for the VNF instance, and contains sufficient information (e.g. IP addresses of VM) to manage and change the deployed VNF instance later on. Figure 5 illustrates an example of VNFR for vOCS.

The STF workflow implementation is organized by VNFC (VNF component). All the VNF related context, including VNFD, VNFR, current VNFC instance, and configuration are available to be accessed as input parameters and can output any result to the context. A STF workflow includes a set of task sequences. It comes with a template that includes STF flows and artifacts. This template is filled and rendered by the different elements that constitute the VNF context, such as VNFC instance, VNFR, VNFD. The typical task types can be remote shell, remote copy to or from. . .

Figure 5 VNFR example for vOCS (key elements only).

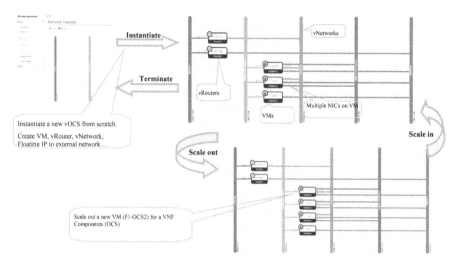

Figure 6   HP Helion dashboard with vOCS operations (Instantiate, Scale in/out, Terminate).

3 Evidence the Solution Works

We have done a PoC (Proof of Concept) with HP OCS, including a vOCS VNF, a VNEM prototype and HP Helion Openstack community edition. Figure 7 illustrates the POC architecture.

The POC (Proof of Concept) environment is built on a HP Helion Openstack platform, running on HP BL blade bare metal servers, and the hypervisor is KVM. The VNEM is running on a Linux VM server. VNEM interacts with HP Helion with Openstack nova and neutron API to create the VM and virtualized networks based on the VNF descriptor, automatically. Once the VNF required virtualized resources are created (instantiated), VNEM interacts with the VMs with SSH based workflow and tasks to execute other operations.

4 Competitive Approaches

NFV is an emerging technology where management and orchestration is the hot topic, but today most of the discussions are focused on the NFV Orchestrator part. The evolution of EM and VNF Manager are up to individual VNF teams, no specifications, no standard and no toolkit or product are available for VNF developers to use. It becomes a major challenge to transform existing Network Function to a Virtualized Network Function and support ETSI NFV MANO (Management and Orchestration). Our VNEM solution facilitates this

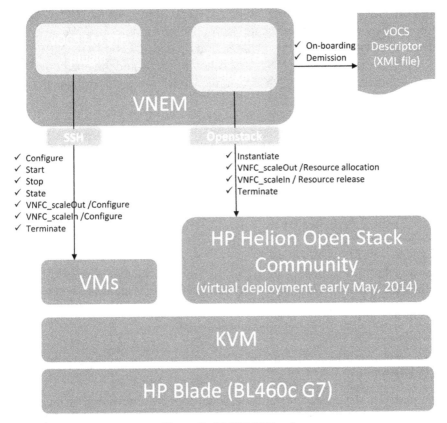

✓ Configure
✓ Start
✓ Stop
✓ State
✓ VNFC_scaleOut /Configure
✓ VNFC_scaleIn /Configure
✓ Terminate

✓ Instantiate
✓ VNFC_scaleOut /Resource allocation
✓ VNFC_scaleIn / Resource release
✓ Terminate

✓ On-boarding
✓ Demission

Figure 7   VNEM POC arch.

transformation and makes it easier and quicker. We have demonstrated the VNEM PoC to many experienced people and got very positive feedback. So far, we are not aware of any Network Function independent solution for this transformation.

5 Current Status

We have implemented a prototype during our PoC, which is illustrated in Figure 7. The Openstack environment is HP Helion virtual deployment edition released in May 2014.

The POC covered all scenarios illustrated in Figure 4. Most of the NFV MANO (Management and Orchestration) scenarios are implemented and demonstrated, for example, automatic deployment with predefined VNFD, create a vOCS within minutes, scale out/in within minutes … We also produced a demo video which is available at: http://pan.baidu.com/s/1sjDgP1z

The video demonstrates the VNF Descriptor, lifecycle and the operations illustrated in Figure 4 with the VNEM CLI, the HP Helion dashboard and the vOCS Component VM interactions.

From overall NFV MANO architecture perspective, VNEM acts as a VNF Manager, and needs to be integrated with an NFV Orchestrator, such as HP NFV Director.

Thanks for the easy and efficient integration approach provided by VNEM, recently, we transformed an HPPCRF (3GPP Policy and Charging Rule Function) product to become a virtualized PCRF (vPCRF) in less than a month. This vPCRF within an integrated vEPC (3GPP Evolved Packet Core) solution was demonstrated in Mobile World Congress 2015. Both vOCS and vPCRF are integrated with HP NFV Director.

We also presented a paper in ICIN15 (The 18th International Conference on Intelligence in Next Generation Networks), and the presentation triggered major interest. Subsequently, in ETSI NFV, new contributions regarding generic VNF Manager were proposed and approved.

6 Next Steps

We have started to share this work and the actual code within our organization so it helps educate other engineering teams on NFV and the work can be reused to accelerate NFV implementations and support in different product lines. This work can be leveraged to productize at Platform level.

Regarding next steps, we are planning to improve the VNEM with more production features, such as:

- VM anti-affinity to support high availability, allowing same VNF Component VM not being created in a same bare metal server to avoid single point HW failure
- A general KPI (Key Performance Index) mechanism described in VNF Descriptor to let VNEM collect real time KPI data to be used for auto scalability and monitoring;
- A scalability decision engine as an embedded component in VNEM.
- A visual VNF instance deployment and monitoring User Interface.
- A VNF Descriptor GUI Editor

- A generic open interface between VNEM and NFV Orchestrator, which is the Or-Vnfm reference point in Figure 1, to enable VNEM be integrated with 3rd party NFV Orchestrator.

References

[1] NFV specifications, ETSI http://www.etsi.org/technologies-clusters/technologies/nfv
[2] HP Helion: http://helion.hpwsportal.com

Biographies

W. Bo is from Communications & Media Solutions (CMS) Business Unit within Hewlett Packard Enterprise (HPE), who addresses the needs and create new business models, deliver new services and improve customer's satisfaction while achieving operational efficiencies for Communication Service Providers (CSP).

As Master Solution and System Architect for the Hewlett Packard Enterprise Services, Bo has more than 19 years of experience in the Telecommunications, IT, Software and Service industry. Bo has provided technical leadership, recommendations and innovation to leading-edge Pan-HP multi-technology products, solutions and deliveries. Bo performs the role of Chief Solution Architect for 2 Products, SNAP (Subscriber, Network and Application Policy) RTC (Real Time Charging) and UPM (Unified Policy Manager), and a Solution Architect for eIUM (Enhanced Internet Usage Manager) Product, driving innovations, visions and strategies for RTC, UPM and eIUM products, HP CMS, pan HP and HP's customers.

O. Marie-Paule is a seasoned HP executive, bring over 25 years of telecom experience. She has deep expertise in both the networking and IT environments, NFV, SDN and M2M/IoT.

Marie-Paule is Distinguished Technologist for HP CMS, Communication and Media Solution organization, focused on customer innovation and emerging trends in the communication industry. She leads the technology discussions for NFV (Network Function Virtualization), M2M, Analytics, Cloud. She seats on ETSI, ATIS, IEEE and other standard bodies. She is Vice-chairman of ETSI NFV ISG, seats on the Technical Steering Committee, is a Vice Chair of TST working group (testing, interoperability and Opensource), rapporteur of SDN work item, and Vice Chair in IEEE SDN. She participates in European Commission SDN-NFV task force, and is also active on M2M & IoT, within ETSI and OneM2M, rapporteur of TG28 low throughput network work items, working with new players such as Sigfox or LORA technologies. She holds a few patents and is also a frequent industry speaker and editor in professional magazines and blogs, incl HP Telecom IQ.

Marie-Paule prior responsibilities include managing HP's worldwide VoIP program, HP's wireless LAN program, and HP's Service Delivery program. Since joining HP in 1987, she has held positions in technical consulting, sales development and marketing in Europe and in the Americas. Those roles have focused on strategic and operational responsibility for Networking, IT and operations in the telecom domain.

Marie-Paule holds a master's degree in Electrical Engineering from Utah State University and business education from INSEAD, Paris. Prior to joining HP, Marie-Paule spent five years with France-Telecom/Orange research and development labs, defining architecture and value-added services launch for corporate customers. She enjoys skiing and outdoors in general.

www.ingramcontent.com/pod-product-compliance
Lightning Source LLC
LaVergne TN
LVHW012332060326
832902LV00011B/1852